CW00521411

in Ireland, 1900-2000

*A century of change in
Irish railways' road and rail
passenger services and vehicles*

Michael Collins

This book is dedicated to the memory of my brother Declan, who shared my interest in transport and who died while this book was in the final stages of preparation.

6 5 4 3 2 1

© M Collins
Newtownards 2000

Designed by Colourpoint Books,
Newtownards
Printed by ColourBooks

ISBN 1 898392 37 4

Colourpoint Books

Unit D5, Ards Business Centre
Jubilee Road
NEWTOWNARDS
County Down
Northern Ireland
BT23 4YH
Tel: (028) 9182 0505
Fax: (028) 9182 1900
E-mail: Info@colourpoint.co.uk
Web-site: www.colourpoint.co.uk

Front cover: *At Strabane in 1960, CIÉ bus No P213, a Leyland OPS3/1 of 1951, is parked across the rails as it loads mail from a train.* WH Montgomery

Back cover: *Translink Volvo B10 in Centrelink livery in Donegall Place Belfast.* Author

The author (right) with the Chairman of Down District Council at the Downpatrick and Ardglass Railway.

Michael Collins, born in 1949, comes from a transport background. His grandfather joined Belfast Corporation as a tram conductor before World War 1 and retired as an inspector in 1947. In the same year, his father joined the NIRTB as a conductor, became a driver and later was an inspector under the UTA and Ulsterbus. As a result, buses and busmen surrounded Michael from his earliest years.

In 1967, whilst a student, Michael's father arranged a holiday job for him as a conductor with Ulsterbus. Michael served as a conductor each summer until 1972. He graduated from Queen's University Belfast in that year with a BA in Geography and Political Science and a post-graduate Diploma in Business Administration. He later upgraded this DBA to an MBA.

From 1972-1974 he was Personal Assistant to Werner Heubeck, Ulsterbus's charismatic Managing Director. During this time he took the opportunity to train as a bus driver in the company's driving school, something Mr Heubeck encouraged all his management staff to do. This later gave Michael the opportunity to drive many of the buses in preservation in Northern Ireland.

His interest in railways also started as a child, encouraged by his father who, during the 1950s, frequently took him to see and travel on trains. In 1986 he joined the newly formed Downpatrick Railway Society and is currently a director and Company Secretary of the Downpatrick and Ardglass Railway. He drives the company's diesel locomotives, is a fireman on their steam engines and is involved in the permanent way construction programme. This book therefore is a synthesis of Michael's interest in road and rail transport, and his training and experience as a geographer and manager.

Michael is currently a Principal Lecturer in management at the Belfast Institute of Further and Higher Education. He is married with four children and lives in Belfast.

Contents

Glossary of Abbreviations

AEC	Associated Equipment Company
AMG	American Motors General
AVM	automatic vehicle monitoring system
BCDR	Belfast and County Down Railway
BCT	Belfast Corporation Tramways
BNCR	Belfast and Northern Counties Railway
BOC	Belfast Omnibus Company
BR	British Railways
BREL	British Rail Engineering Limited
CDR	County Donegal Railways (abbreviation of below)
CDRJC	County Donegal Railways Joint Committee
CIÉ	Córas Iompair Éireann
CLR	Cavan and Leitrim Railway
CME	chief mechanical engineer
CTC	centralised traffic control
CVBT	Castlederg and Victoria Bridge Tramway
CVR	Clogher Valley Railway
DART	Dublin Area Rapid Transit
DBST	Dublin and Blessington Steam Tramway
DKR	Dublin and Kingstown Railway
DMU	diesel multiple unit
DNGR	Dundalk Newry and Greenore Railway
DSER	Dublin and South Eastern Railway
DTA	Dublin Transportation Authority
DTI	Dublin Transportation Initiative
DUTC	Dublin United Tramways Company
DVLR	Derwent Valley Light Railway
DWWR	Dublin Wicklow and Wexford Railway
EMU	electric multiple unit
GAC	General Automotive Corporation
GEC	General Electrical Company
GM	General Motors
GNR	Great Northern Railway (Ireland)
GNRB	Great Northern Railway Board 1953-58
GOC	General Omnibus Company
GSR	Great Southern Railways
GSWR	Great Southern and Western Railway
GWR	Great Western Railway
IÉ	Iarnród Éireann
IOC	Irish Omnibus Company
IRRS	Irish Railway Record Society
ITT	Irish Transport Trust
LER	Londonderry and Enniskillen Railway
LEV	light experimental vehicle
LHB	Link Hoffman Busch
LLSR	Londonderry and Lough Swilly Railway
LMS	London Midland and Scottish Railway
LNER	London and North Eastern Railway
LNWR	London and North Western Railway
LRO	light rail order
LRT	light rapid transit
MCW	Metro-Cammell Weymann
MED	multi-engined diesel
MGWR	Midland Great Western Railway
MPD	multi-purpose diesel
MR	Midland Railway
NESC	National Economic and Social Council
NCC	Northern Counties Committee
NIR	Northern Ireland Railways
NIRTB	Northern Ireland Road Transport Board
NITHC	Northern Ireland Transport Holding Company
OMO	one man operation
OPO	one person operation
PFI	Private Finance Initiative
PSV	public service vehicle
QBC	quality bus corridor
SLNCR	Sligo Leitrim and Northern Counties Railway
TMSI	Transport Museum Society of Ireland
TVO	tractor vapour oil
UFMT	Ulster Folk and Transport Museum
UTA	Ulster Transport Authority
UTR	Ulster Transport Railways

lbs	pounds (weight)
sq ft	square feet

Introduction

For over a century, a feature of railway passenger-carrying operations in Ireland has been the use of trams, steam railmotors, railcars, buses and bus-type rail vehicles as part of the service provision of railway companies. Some companies adopted them, to use modern management jargon, proactively, with enthusiasm, and made use of them widely as an integral part of their provision; others used them often reactively, with less enthusiasm, as marginal, experimental and often short-lived features of their passenger carrying provision.

The operational reasons for introducing these vehicles varied from company to company and changed over time as technology and economic circumstances changed. Consequently, the period covered by this book can be roughly divided into four phases.

The first phase covers the early years of the twentieth century, prior to the outbreak of World War I. The second stretches from the end of World War I to about 1948. Phase three covers the period from 1948 until about 1980. The final, fourth, phase is in its early years, but some of its aspects are becoming clearer now that we are at the millennium.

During the first phase, which really has its origins in the 19th century, road transport was not seen as any real competition to the railways. For example, at the dawn of the railway age, it took 12-13 hours, spread over two days travelling by mail coach over rough roads, to make the journey from Belfast to Dublin. In 1896, when the Speaker in the House of Commons suggested that one day the motor car might rival light railways, he was laughed at. In fact, the entire network of road services had collapsed relatively quickly as the railways spread throughout the country after 1850, although, even at this time, some road services had been adapted as feeders to the railways.

A major reason for the success of the railways during the 19th century was that the horse-drawn transport technology used on the road system was so poor. This in turn led to the deterioration of the road system, which was the responsibility of either the County Grand Juries, the main road-building agencies in the country, or private sector Turnpike Trusts approved by statute.

The latter had been appointed by Parliament, in the first half of the 18th century, to maintain and control a relatively small number of the most important roads in the country. They had permission to meet the expenditure thus incurred by collecting tolls from all road users at turnpikes.

In the first half of the 19th century, the Grand Juries inherited the turnpike roads as their traffic declined under the impact of railway competition. The new railways often closely paralleled the turnpike roads. The last turnpike, which was between Antrim and Coleraine, closed in 1858. In 1898 all road building and maintenance responsibilities were passed to County Councils under the Local Government (Ireland) Act.

Specialised rail systems, such as steam or horse-powered tramways, were established and operated during the latter part of the 19th century by several railway companies. These systems were often extensions to their main rail networks, where the provision of a full-scale rail line was thought, for one reason or another, to be unsuitable. Sometimes the traffic and/or the distance did not justify the expense of building or providing a full-scale steam operated train service. This, for example, was the case with the Fintona horse tram and the Portstewart Tramway. Sometimes the topography was too severe for the operation of steam trains, as was the case with the Hill of Howth electric tramway near Dublin.

Just prior to World War I, railway companies began to introduce specialised rail vehicles. These could be operated at lower costs than traditional locomotive-hauled trains in order to meet actual or anticipated competition from road vehicles. We see this with the introduction of steam railmotors in the early 1900s. Railmotors were designed to develop traffic emanating from the spreading suburbs and to block competition from the new urban tramway systems, or sometimes as a cheaper way of operating

MR (NCC) Steam tram at Culmore station waiting to depart for Portstewart.   CP Friel collection

poorly patronised branch lines.

The second phase, beginning after World War I, finds the railways having to deal with increasingly intensive competition for both passenger and freight traffic. This came from a largely unregulated and rapidly growing road transport industry. This sudden growth in road transport was caused initially by a flood of army surplus internal combustion engined vehicles, and men capable of operating them, released into the country after World War I.

The introduction of regular motorised road services, after the war, highlighted the neglect which the country's roads had suffered since the late 19th century. As a result, local government began once more to spend money on road improvements.

Initially, these road vehicles were not seen as a problem by the railways. For example, in 1922 the Northern Ireland Light Railway Commission, set up to examine the future of such railways in the new Northern Ireland state, concluded that, "motor transport would become an auxiliary rather than a competing factor to be economically reckoned with."

However, by the end of the period under discussion, this point of view had been shown to

have been way off the mark. All of the railway companies were in serious financial straits as the result of road competition. They started to ask the governments in both Dublin and Belfast for help in meeting their financial obligations and for legislation to protect them against what they regarded as unfair road competition.

During the 1920s numbers of a new type of passenger vehicle were deployed by the railways to try and counter road competition. Steam railcars, conceptually very similar to those introduced before World War I, were introduced by several Irish railway companies. A sign of the changing times was that these vehicles were promoted by their builders specifically on the basis of their favourable operating economics vis-á-vis road bus operations. This was a marketing pitch which would be used again in the future with diesel traction.

From the 1920s to the 1940s, following the steam railcars, purpose-built single-unit light-weight petrol or diesel railcars were introduced by several railway companies, utilising bus engine and bodywork technologies. Most of these railcars were diesel powered, with mechanical drives, although

A 32-seat Commer NF6 belonging to the Imperial Bus Co. This bus later became NIRTB H437.   Author's collection

LMS (NCC) railcar No 4 at York Road station in 1965.   Roger Holmes

one must not forget the Drumm battery-powered electric trains of the Great Southern Railways. These were successful enough, although not replicated outside their original operating area. This period also saw the provision of railbuses, converted from road buses on very lightly trafficked routes.

Nevertheless, for most of this period, all of these vehicles were marginal, and sometimes experimental, parts of their owners' fleets, the main work, as before, being carried out by conventional steam traction.

The third phase of this story, beginning just after World War II, finds the railways attempting to achieve the maximum operational economies possible at a time when it was clear that they had essentially lost the war with road transport. They were now having to find new niches for themselves in passenger and freight transport markets, usually after being amalgamated under direct or indirect state ownership, and on ever-shrinking route networks.

The evolution of rail vehicles during this period saw, in the 1940s and 1950s, the introduction of diesel powered, multiple unit railcars, again using bus engine technology. These were founded on the success of the earlier single unit diesel railcars. The development of multiple unit vehicles became possible as vehicle control and gearbox technology improved.

The superior operating economics of diesel and later diesel-electric vehicles, compared with steam locomotives, coupled with increasingly intensive road competition, resulted in diesel traction rapidly displacing steam in all parts of the Irish railway network.

In the late 1960s trains were developed, hauled by one or more power-cars, which were a combination of a diesel-electric locomotive and a passenger carriage. Such diesel-electric sets have become the standard configuration for all but the heaviest long distance passenger trains on Northern Ireland Railways (NIR). Long distance NIR trains are locomotive hauled.

Until the introduction of the Dublin Area Rapid Transit (DART) electric trains and, more recently the new Mitsui and GEC 'Arrow' railcars, Iarnród Éireann (IÉ) and its predecessor, Córas Iompair Éireann (CIÉ), had preferred to use locomotives

even for short haul and suburban services.

From the mid 1920s, in addition to the development of rail vehicles to improve traffic figures and operating economics, many railway company managements decided, with regard to road operations, to adopt the old motto 'If you can't beat them, join them'. As a result, they began to introduce their own bus services.

At first, the use of buses was often designed to eliminate private road competition, or to feed the railways with passengers from areas where it would not have been feasible to build a branch line. Some railway companies later became major bus operators in their own right. They provided bus services which paralleled or replaced part of their rail provision. In some cases, this led ultimately to complete replacement of rail passenger services.

Railway-owned buses were introduced in a small way around the turn of the century but the main development of railway bus fleets was in the 1920s and 1930s. Changes in the law allowed railway companies to buy up private bus operators who were competing with the rail services and then develop bus services as feeders to the railway, or as additional provision to their rail services.

Where the larger railway companies operated road bus services, they often built and maintained their bus fleets using the same facilities as they used for their rail vehicles. They also operated through-ticketing systems and common road-rail timetables. In Northern Ireland, this sensible pattern of development was brought to an end by government intervention, though this was not really the intention. However, it survived longer in Éire.

Recent years have seen the reintroduction in Éire of the diesel multiple unit configuration for suburban services, a move away from the diesel-electric locomotive-hauled train for this type of work. The mid-1980s saw the introduction of electric traction into Irish rail transport, using a system picking up current from overhead wires. This is now the basis of the very successful Dublin Area Rapid Transit or DART, which has attracted much traffic back to the railway in the Dublin Bay area.

In addition, early in 1995, Northern Ireland Railways announced that they were going to replace their 80-class diesel-electric units, introduced in the

LLSR Reo FBX fleet No 29, acquired from J Winter, Derry, in 1931. Author's collection

mid-1970s, with diesel multiple units similar to the IÉ 'Arrow' sets. However, it is not yet clear when these new NIR units will enter traffic.

It is also evident that there is growing public unease about the problems caused by the growth in road traffic and the consequent road building programmes.

In the 1980s there was pressure to re-introduce electric tramways into Belfast and Dublin. These would be capable of operating both on street and over lightly trafficked or re-opened suburban rail lines, as has begun to happen in parts of Britain over the past few years.

At the time of writing, such a tramway or Light Rapid Transit system (LRT) is to be introduced into Dublin, the first phase to be completed early in the new century. In addition, increased investment in the conventional railway infrastructure and in new locomotives and rolling stock is now taking place in both Northern Ireland and the Republic of Ireland. Belfast's city centre railway terminus, Great Victoria Street, which was closed in 1976, was rebuilt and re-opened in late 1995.

Developments and changes also continue apace in the management structures of the island's two main transport companies.

In 1987 CIÉ, which had operated as an integrated transport company since its foundation, was split up into three main operating companies, Irish Rail, Dublin Bus and Irish Bus. CIÉ remains as a holding company to provide a level of overall co-ordination between the operating subsidiaries.

In 1995 an important announcement was made by the British government with regard to the organisation of public transport provision in Northern Ireland. The Minister of State announced, on 16 January 1995, that in future the emphasis on road building would be replaced by a greater emphasis on the development of public transport.

To facilitate this change a new corporate structure has been set up to integrate the operations of Northern Ireland Railways, Ulsterbus and Citybus. It is charged with the setting up of an integrated public transport provision for the province with a common timetable and through-ticketing. It has adopted the name 'Translink' as a common brand name for all road and rail services, although the previous corporate names will be retained.

In making these changes, CIÉ seems to have

adopted an organisational structure closer to that just abandoned in Northern Ireland, and the new structure being adopted by 'Translink' is closer to the organisational structure which CIÉ has recently decided to give up!

In a further move, the Department of the Environment, Roads Division, responsible for road building in Northern Ireland, has been merged with its Transport Division, responsible for overseeing public transport provision. A contentious proposal to build a major road through a suburban beauty spot in Belfast has also been cancelled and Ulsterbus has proposed the establishment of a guided busway to relieve the traffic pressure in this area.

The negative effects of 70 years of the unbridled expansion of road transportation are now really starting to make themselves felt. Increasing traffic congestion and environmental pollution, particularly in cities, are generating additional costs for industry and society. Changes, such as those outlined above, would indicate that a new phase in the history of public transport, road and rail, may now be imminent.

This book is intended to provide in one volume, drawing mainly from existing literature, a brief survey of the evolution of passenger traffic provision, in Ireland, during the twentieth century. It will also deal with the associated development and use of rail vehicles which evolved primarily as the result of road competition and which in their later stages of development incorporated road vehicle technologies. The history of the use of road passenger vehicles owned or operated by railway companies will also be covered.

By and large tramways are not included, except in a couple of cases where vehicles introduced by tramways are important to the mainstream of overall vehicle development. However, the distinction between a 'tramway' and a 'railway' can be somewhat arbitrary. For example, the Clogher Valley system is covered mainly because of the importance of the diesel traction it introduced. However, in its latter years it called itself a 'railway', although for most of its route it was a roadside tramway.

Neither does the book deal with normal steam-hauled trains, diesel multiple unit trains nor with the diesel electric powered trains introduced in the late 1960s by the Ulster Transport Authority and later by Northern Ireland Railways. It is felt that these vehicles are the result of mainstream railway technology rather than hybrid road/rail technologies and their history is comprehensively covered elsewhere in the literature on Irish railways.

It is not intended to cover in detail the activities of purely road transport operators. However, in order to give as complete a context as possible to railway developments from the 1920s onwards, brief histories of the establishment and evolution of the major integrated transport companies – the Irish Omnibus Company, the Great Southern Railways, the Northern Ireland Road Transport Board, the Ulster Transport Authority, Córas Iompair Éireann and Translink – have been included.

Chapter 1

The emergence of road/rail competition
1900-1945

Introduction

This brief survey deals with the establishment and development of the GSR, the IOC and NIRTB. It shows how the governments, as well as the managers, of the public transport companies in Ireland's two states had to wrestle with the problems of providing efficient and effective public transport systems for their jurisdictions, while coping with the problems of financial deficits and political sensitivities which are inevitably part of the provision of such services.

The survey is also important in chronicling how, in the face of growing competition from road transport, governments and state transport providers developed very different attitudes to the maintenance of a railway network, and attempted to strike the 'correct' balance between provision by road and rail.

1900-22: The years before the political division

Between the dawn of the railway age in Ireland, in 1834, and 1925, when most railways in the new Irish Free State were amalgamated into the Great Southern Railways, about 3,000 miles of railway were constructed. It had been hoped by their various promoters that these railways would help develop Ireland economically but, by the 1860s, it was being suggested in influential circles that things could be improved if there were fewer companies operating the country's railway system.

In 1872, in a joint report of the two Houses of Parliament, an argument was made for state ownership of the railways. During the following decade a number of attempts were made to implement this idea.

In 1910 a report from a Viceregal Commission recommended that "an Irish authority be instituted to acquire the Irish railways and work them as a single unit". It also recommended that an annual grant be paid by government to the new railway authority. This radical recommendation was the subject of the Commission's majority report, but even the minority report recommended that an amalgamation should take place and that, if this did not happen voluntarily, that it be forced on the railway companies. The

Commission believed that amalgamation would lead to a more efficient railway operation and so to an improved service to the public. Fifteen more years passed before such a merger took place.

The main railway companies were not unhappy in principle about the idea of such a merger, but they saw difficulties regarding ascertaining the compensation to be paid to shareholders. By the outbreak of World War I, still nothing had been resolved.

In 1916 the state took control of all Irish railways. Subsequently costs tripled, but revenues only doubled. Thus, the seeds for future financial disaster were sown.

After World War I, private bus operators began to spring up all over the country. Competition between these unregulated operators and between them and the railways was intense, and the railways began to suffer significant financial damage.

In 1921 Ireland was divided and in 1922 the south was given independence as the Irish Free State. Northern Ireland remained within the United Kingdom. From that time, the patterns of development of Irish railways, north and south of the new border, began to diverge.

The Irish Free State 1922-45: the GSR and IOC

During the Civil War which followed independence, the Dublin government continued to control the railways but, as in the recent 'troubles' in Northern Ireland, the railways were frequently the target of hostile armed action..

In the spring of 1922, the government set up a Railways Commission and its report later that year recommended state ownership. In January 1923 the government announced that if the railways did not merge voluntarily by July, legislation would be passed to force them to do so. This threat had no effect.

In the same month, the largest railway company in the Free State, the Great Southern and Western, announced that because of financial difficulties, it intended to cease trading. The government provided financial help so that the GSWR could keep going. Other railways in the new state were experiencing similar difficulties so matters could only get worse.

Foundation of the GSR and IOC, 1925-26

In 1924 the Irish government introduced a Railway Bill which provided for the merging, initially, of the GSWR, the Midland Great Western Railway and the Dublin South Eastern Railway into a new Great Southern Railways. The GSR would then absorb the 22 other railway companies in the state. The GSR was formed on 1 January 1925.

In the years which followed, it became clear that the financial viability of the GSR was being threatened by the growth in road transport. When the GSR had been founded there had been warnings of the potential threat from road transport, but it was thought that once the improvements from organisational efficiencies began to appear, the railways would be able to compete successfully. However, railway traffic and revenues began to fall and the GSR lobbied to be allowed to compete on equal terms with road operators who were subject to almost no restrictions.

In 1926 the Irish Omnibus Company (IOC) was incorporated with the objective of developing bus services in the Dublin area. The following year, the Railways (Road Motor Services) Act was passed which empowered the railways to operate buses. Following this, the GSR and the IOC reached an agreement under which the IOC took over the GSR's road licenses.

The IOC began to operate throughout the Irish Free State and where possible either bought out the competition or ran it out of business. It also took over the GSR's charabanc and coach services which had

IOC Leyland LT4. This bus later became part of the Erne Bus Service fleet and ended its days as UTA K206 in 1960.

Author's collection

Former MGWR Broadstone station in Dublin with GSR buses.   National Railway Museum

been inherited from some of the GSR's constituent companies.

In 1929 the GSR took a controlling interest in the IOC but was still unhappy with its position vis-à-vis road passenger and freight competition. In 1931 the GSR chairman called for fresh legislation "regulating transport and removing the disabilities at present imposed on railways".

By that time, the number of licensed road vehicles had increased to 49,000 compared to 29,000 when the GSR was set up. All forms of road transport were virtually unregulated and the situation was getting seriously out of hand.

In 1932 the government introduced a Road Transport Act which required all scheduled passenger services to be operated under licenses issued by the state. Timetables and charges would also have to be published. Independent operators could continue in business, provided they met the requirements of the Act.

The following year, a further Road Transport Act was passed which gave the railways the power to compulsorily purchase their passenger and freight road competitors. This legislation was designed to restore a monopoly in each area so that there would be one efficient transport provider. It was also intended to prevent the under-utilisation of the railways which the government regarded as an important national asset.

As a result of these Acts, four operators – the GSR, the Great Northern Railway, the Londonderry & Lough Swilly Railway and the Dublin United Tramway Co – ended up controlling most of the bus services within the Irish Free State.

In 1934 the GSR absorbed the IOC, which was then wound up. Its assets became the Omnibus Department of the GSR.

In the 1930s, private road transport grew rapidly. The number of private cars rose from about 9,000 in 1923 to just over 48,000 in 1938. However, independent bus companies fell in number due to the acquisitions being made by the IOC/GSR and the DUTC.

During the 1930s, the GSR also began to close railway lines. The first to go, in 1932, was the Cork to Crosshaven narrow gauge line. In 1934 the Bagnelstown to Palace East and the Edenderry, Castlecomer, Killala and Killaloe branches lost their passenger services. The Kinsale branch was closed completely, as was the Cork to Muskerry narrow gauge line. In 1935 the Clifden line was closed and September 1937 saw the demise of the Westport to Achill line. In 1939 the Tralee to Dingle line became goods only, whilst its Castlegregory branch was closed completely. In these cases, GSR buses replaced the trains.

The Ingram Report 1939

In spite of these rationalisations of the railway network, problems continued to mount. Traffic fell and finances deteriorated. As a result, in 1938, the government appointed a Tribunal of Inquiry on Public Transport under the chairmanship of Mr John Ingram. He had been secretary to the 1922 Railway Commission, the report of which had led to the creation of the GSR.

The Tribunal's majority recommendation was that a National Transport Council be set up for the review of all forms of transport. The report also set out the problems familiar to students of Irish railways during this period:

Railways must accept all and any class of traffic, operate on the basis of published timetables irrespective of fluctuations in demand, provide a permanent way and maintain it, must provide for safety of passengers and goods by operation of their own signalling systems.

At this point, a significant divergence in attitude between people of influence north and south of the border becomes obvious. The GSR had put forward a proposal that it withdraw train services on about one third of its network and replace them with buses. It also proposed to build four new bus stations. The Tribunal did not like this idea at all and the report spoke of the fundamental importance of the railway's role in handling peak traffic and as a stand-by for privately owned transport. They recommended increased duties on private road transport with the money used to fund the railways. In effect, they proposed cross-subsidisation between private road and public rail services.

The Tribunal's minority report, written by an economist, Dr Kennedy, recommended state ownership and subsidies rather than restrictions placed on private road transport, which he believed would have a detrimental effect on national economic development. In 1950 Kennedy's recommendation would, in effect, be implemented with the setting up of CIÉ Mark 2.

Within a week of the publication of the two reports, World War II broke out. Shortage of fuel provided a respite for both the GSR and GNR. Severe petrol rationing forced traffic back on to the railways, which again were placed under government control. However, as the 'Emergency' progressed the quality of the service that the GSR provided gradually deteriorated due to coal shortages and the poor quality of much of the fuel that was available.

In 1942, 11 more branch lines were closed. Many others had no services and mainline services operated only two days per week. The GSR's operational problems grew, but its financial problems decreased.

In 1942 the government appointed a new chairman, Percy Reynolds, to the GSR. He immediately started to apply pressure on the government to take action to make the railways viable. As a result, in 1944 the Transport No 1 Bill was published. This proposed to wind up the GSR and establish a new company to be called Córas Iompair Éireann. The Bill became the 1944 Transport Act and CIÉ was established in January 1945.

At the time of its absorption into CIÉ, in January 1945, the GSR worked all the main provincial bus services south of a line between Dublin and Sligo, those north of this line being operated by the GNR. This was as the result of agreements entered into in 1932 between the IOC and the GNR and between the IOC and the DUTC. The GSR also operated the Cork, Waterford and Limerick city bus services.

Northern Ireland 1921-1945: the NIRTB

A major part of this story involves the establishment by the Northern Ireland government, in 1935, of the Northern Ireland Road Transport Board and, in 1948, its successor, the state-owned Ulster Transport Authority. I feel, therefore, that a brief history of these organisations is essential to the understanding of the interrelated nature of road and rail passenger transport in the province from 1935 until the late 1960s.

Private sector competition 1921-35

At its foundation, the new state of Northern Ireland had about 800 miles of railway, operated by 13 companies. About 75% of the mileage was operated by three companies – the GNR, the LMS (NCC) and the BCDR. Five companies operated on both sides of the new border but only one, the GNR, had a substantial mileage on both sides.

In the period between 1921 and 1935, the

Belfast Omnibus Company vehicles lined up outside the Custom House, Dublin.   Frank Byrne collection

HMS Catherwood showing the capabilities of a new low bridge double-decker at Derriaghy railway bridge.
Ribble Enthusiast's Club

financial position of the railways operating within Northern Ireland had become increasingly precarious, as the result of growing road competition following World War I. In fact, the real beginnings of this competition coincided with the formation of the state of Northern Ireland in 1921.

In that year, there were over 11,700 locally registered motor vehicles in the province. Half of these were motor cycles, 3,382 were private cars and 1,607 hackney carriages. In this latter figure, no distinction was made between buses and taxis, so the majority of vehicles in this category were probably taxis. By 1925 the number of buses or charabancs had risen to 400 and by 1934 this number had again risen to 677, while the number of private cars had risen to about 48,000.

The initial political problem for the new government in Belfast was in establishing, for the purposes of regulation, under which jurisdiction the cross-border railway lines lay. Up until 1921, as with all the railways in the UK, such regulation had been the responsibility of the Westminster government.

Under the Government of Ireland Act, by which both Northern Ireland and the Irish Free State had been established, regulation of the railways was to have been the responsibility of a Council of Ireland. However, this Council had not been set up, and as a temporary measure the Dublin government had been granted powers of supervision over the country's railways.

As it was rumoured that the Dublin government intended to nationalise the railways in the Irish Free State, the Belfast government was afraid that the Northern Ireland portions of cross-border railways, and in particular the GNR, would come under the control of the Dublin government.

In 1922 Belfast received assurances from London that this would not be the case and, in a letter reporting this fact to the GNR's Belfast directors, Northern Ireland's Prime Minister, Sir James Craig, referring to British legislation, urged the railway company managements to investigate the possibilities of merging their operations as a way of strengthening their financial situation:

> The Northern government are disposed to think that the advantages of the (British) Railways Act 1921, as regards grouping ... should be made available to Northern Ireland; and, if the representatives of the railways in Northern Ireland should also favour this

course, the government would urge the Imperial government (Westminster) to take the necessary steps for legislation...

Craig went on to make a further proposal which, if taken up, would have totally changed the future of the province's railways:

> ...[the government] would further desire you to be aware that, if there was a possibility of fusion with the London Midland and Scottish group in Great Britain, they would unhesitatingly welcome the proposal and give any assistance in their power towards bringing the negotiations to a successful conclusion.

As a temporary measure, jurisdiction over the railways in Northern Ireland was placed under the control of the Ministry of Transport in London. When it had become clear that the Council of Ireland would not be set up, responsibility was transferred, in 1925, to Belfast.

No action on Craig's suggestion was taken by the managements of the province's railways.

The Railways Commission 1922

In May 1922 a commission was set up by the Northern Ireland government. Its task, set out in the terms of reference, was:

> ... to advise the Government of Northern Ireland as to what changes, if any, are desirable in the administration of the railway undertakings in Northern Ireland and in particular to report on
>
> • The financial position and earning power of the various railway companies.
>
> • The best means of consolidating or otherwise working the different railways.
>
> • The remuneration and conditions of employment of the various staffs.
>
> • Any other relevant matters.

The Commission issued two reports. The majority report recommending, *inter alia*,

• Continuance of the present "well-established and competitive system" of private management.

• Absorption of the narrow gauge Clogher Valley Railway by the GNR and of the narrow gauge Ballycastle Railway by the LMS (NCC).

• Greater fluidity of movement of rolling stock, locomotives etc, between the various companies.

The minority report, however, recommended

nationalisation of the entire rail network in Northern Ireland and pointed out the wastefulness of a system by which 800 miles of railway were controlled by 13 different companies. Very little changed as a result of the Commission's work.

As competition intensified, and feeling the financial pressures building up, the railways lobbied the government for assistance. Their preferred options were either that they acquire a monopoly of all road-bourne public transport (buses and freight), or that road operators be required to carry the full costs of the roads they used and be subject to the same stringent operating and wages legislation as the railways – in other words, the creation of a 'level playing field' for the two transport modes.

The new government was wary of the possible political implications of either of these courses of action. They feared a possible electoral backlash from a population which might feel that the big combines were being favoured at the expense of the common man since the flexibility and cheapness of road transport were obvious boons to the ordinary citizen.

The Pole Committee, 1934

As the financial position of the of the railways continued to deteriorate, the Northern Ireland government called in Sir Felix Pole, ex-General Manager of the Great Western Railway in England, to head a committee to look into the whole problem of transport in the province and suggest a solution. His terms of reference called for a policy which would "be fair to the interests concerned and also ensure an efficient system of transport for Northern Ireland".

Pole's basic position was that no matter how good or extensive the road services operating in the province were, they could never satisfactorily replace the railway services. In his eventual report he gave the following warning:

> ... if transport by road and rail continues to be conducted on the present unsatisfactory basis, serious results will follow and ... the people of Northern Ireland will find themselves in the position of having an inadequate and expensive transport system.

That prediction may have taken half-a-century or so to come to pass, but come to pass it certainly has.

Pole's first preference was for the setting up of a Northern Ireland Transport Board, but he did not recommend this because of the difficulty of including

all the province's railways, a number of which, as we have seen, crossed the international frontier. The railways, in their evidence, had not favoured such an amalgamation. By contrast, the two largest bus companies, HMS Catherwood and the Belfast Omnibus Company (BOC), were keen on the sort of board that Pole had in mind, provided it was linked with the closure of branch lines and 'circuitous' railways.

What Pole eventually proposed was the setting up, by the government, of a Road Transport Board. This would compulsorily acquire and run all road passenger and freight operations in the province and operate in conjunction with the railways to provide an integrated system of transport for Northern Ireland. The railway companies should be allowed to invest in it and there was to be a pooling system whereby the railways were to be allocated a proportion of the profits of the Board, after interest had been paid on government loans and on the stock issued as compensation to previous owners which, of course, included the railway companies themselves.

The railways supported this proposal, even though it meant losing their bus and lorry fleets to this new organisation. Pole had hoped that the new Board would include the Belfast Corporation Transport Department's fleet, but the Corporation successfully lobbied to be excluded from the scheme.

The NIRTB, 1935-48

The Northern Ireland government promptly accepted Pole's plan, happy to see a way out of this political minefield. The necessary legislation was passed and the Northern Ireland Road Transport Board came into existence in 1935.

The members of the Board were to be appointed by the Ministry of Home Affairs. It was to be financed by the issue of stock, subject to the approval of the Ministry of Finance, and was required to issue stock to the owners of the undertakings it bought over for so much of the purchase price as exceeded £5,000. It was required to acquire "every road motor under-taking operated for hire or reward ... including the passenger and freight road services of any railway company ..." There were certain exceptions. In particular, the LLSR fleet was exempt, as its sphere of activities lay largely in the Irish Free State.

In order to co-ordinate their undertakings as

NIRTB Leyland TD4. Note the early 'wheel and bar' totem and the conductor's 'Willybrew' punch.

Frank Byrne collection

required by the Act, a standing joint committee of six representatives was set up, three from the NIRTB; one from the GNR, who would also represent the Sligo Leitrim and Northern Counties Railway; one from the Northern Counties Committee, and one from the Belfast and County Down Railway. This committee was charged with preparing a pooling scheme as required by the Act.

The NIRTB was duly established and on 1 October 1935 took over the bus operations of the 'big five' – the BOC, Catherwood, the GNR, the LMS (NCC) and the BCDR. Another 57 smaller undertakings were taken over by the end of the year.

Three smaller operators were left "for special reasons to be dealt with at as later stage". The first was Cassidy's Erne Bus Service of Enniskillen, which had been founded in 1929 and had built up a considerable business on both sides of the border. The second was Hezekiah Appleby's Central Bus Service, also based in Enniskillen, with its main route to Sligo. Finally, there was The Major bus service owned by WJ Clements, which continued to operate in Belfast under an agreement with Belfast Corporation.

The Board took over a fleet of 692 buses of 27 different makes, many of them in need of replacement. Freight acquisitions followed the take-over of the bus services.

From the very beginning, the Board was unable to meet even the interest charges on its government loans and so was unable to pay dividend on the shares issued, much to the anger of the railway managements. In fact, the railway companies found that they had to contribute substantial sums towards maintaining the Board's road transport system which was unable to realise a profit on its own services.

Payment to the 'big five' had been made entirely in the Board's own stock, while most of the smaller operators were paid in cash. The formula for valuing the 'big five's' assets had been worked out by the Ministry of Home Affairs as follows:

- The written down value of the tangible assets at the date of transfer.
- A sum for goodwill, ascertained by multiplying by ten the balance of profits remaining after deducting 4% of the value of the assets

ascertained under the first.

The Board interpreted the Act as requiring the use of this formula for all the undertakings it acquired.

The financial results of this decision are set out below. Note in particular the value of 'goodwill'.

Table 1: Financial Background to the Establishment of the NIRTB passenger undertaking.

A **Consideration paid for acquired undertaking**	**£**	**£**
The big five:		
Belfast Omnibus Co	390,291	
HMS Catherwood	100,395	
BCDR	18,792	
GNR	39,763	
LMS (NCC)	170,381	719,622
Selected larger operators:		
Imperial Bus Services	40,000	
WE Hobson	36,081	
J McCartney	23,464	
W Irvine and Sons	21,000	120,482
Remaining operators:		2,158,111
Total		**2,998,215**

B **Breakdown of consideration payments**	**£**	**£**
Tangible Assets:		
Omnibuses and coaches	267,302	
Other road vehicles	241,914	
Land and buildings	152,141	
Plant, machinery, furniture, etc.	33248	694,605
Intangible Assets:		
Goodwill	*2,257,309*	
Compensation and Severance	46,301	2,303,610
Total		**2,998,215**

C **Breakdown of annual receipts and expenses of acquired undertakings:**

	Big five £	Others £	Total £
Gross Traffic Receipts	521,316	166,858	688,174
Less Operating Expenses (incl depreciation)	486,278	134,561	620,839
Net Profit	**35,038**	**32,297**	**67,335**

(Note: It was on the basis of the figures in Part **C** above that the consideration payments shown in Part **B** of the table were calculated.)
Source: Report of the Committee of Inquiry (Cmnd 198 of 1938); Annual Reports of the Board.

In order to replace many of the life-expired vehicles inherited from the companies it had absorbed, the NIRTB bought 121 new buses in its first year. This expenditure added further financial pressures to the new undertaking.In fact, the Board made an overall loss on its first three year's activities. The loss on its passenger services was £20,713 in 1936, followed by a small profit in 1937 which set the trend for the rest of the pre-war period. However, it lost money on its freight operations up to 1940.

The NIRTB was over-capitalised and nothing that it did in the remaining years before the war improved its position. Another major problem was that the membership of the Board included no-one with transport experience and it suffered from a shortage of skilled technical and clerical staff.

From the outset another source of friction began to develop between the NIRTB and the managements of the railway companies. The latter began to claim, increasingly vociferously, that instead of co-ordinating transport between road and rail, the Board was, in fact, increasingly competing directly with the railways. They said, in effect, that all the establishment of the NIRTB had done was to create a streamlined road transport competitor and that the railway companies were now in a worse position than they had been before the Board was formed.

So great was the gap between Sir Felix Pole's intentions and reality during the NIRTB's first year of trading, that following the publication of the Board's first Annual Report in the summer of 1936, a conference took place between representatives of the Northern Ireland Government and the railway companies. The purpose of this conference was to examine the problems associated with the implementation of the Act creating the Board.

Several important criticisms were made. Firstly, there were numerous public complaints about the reduction of transport facilities and services, and the increases in fares. Secondly, complaints had been received from former owners of transport undertakings about delays in receiving compensation and hearing appeals. Thirdly, complaints had been made by recipients of compensation with regard to the method of payment. A deputation was also sent by several of the standard gauge railway companies, to the Minister of Home Affairs, to emphasise the unsatisfactory nature of the pooling arrangements.

Further investigations into transport, 1938-39

By 1937 it was clear that the NIRTB was not fulfilling the purposes for which it had been set up and the Government decided to have a further examination into the whole structure of public transport in Northern Ireland. In 1938 and 1939 three separate bodies met to again try and sort out the whole complex problem. These bodies were:

- An open fact-finding inquiry under the chairmanship of the Recorder of Belfast, Judge Thompson.
- A private inquiry by a small committee under the chairmanship of Sir William McClintock.
- A Joint Select Committee of the two Houses of the Northern Ireland Parliament which was set up to consider the findings of the first two inquiries and formulate its own opinions.

In brief, three main recommendations were put forward by these three inquiries:

- The continuance of the principal rail services was in the public interest.
- Road services should remain in the public domain.
- Railways should be assisted from road service revenue.

However, there was no agreement on how these objectives should be achieved. The basic problem identified by Judge Thompson's report was that it was plain that the partnership between the Board and the railways had been a failure from the start and their joint committee on 'pooling' arrangements had carried out much discussion but with little practical outcome.

This was the result of a 'fundamental divergence' of views. The railways had seen the 1935 Act as a way of stemming the loss of their traffic to the roads, while the Board was concerned first and foremost with the development of its own organisation.

The McClintock Committee had criticised the Minister for not appointing anyone with transport experience to the Board. It also remarked on the lack of experience of compulsory acquisition which had caused mistakes in the early months of the Board's existence. It noted that Sir Felix Pole had envisaged a slower process of acquisition, based on a merger of the main undertakings followed by purchase of the rest. The committee did not believe that it had been the proper time for the acquisition of the freight undertakings at all and that it would have been a better policy for the government to have regulated the industry with a view to encouraging amalgamations, and then for the Board to have taken over the larger groups which would have emerged.

The McClintock Committee also felt that the payments for goodwill were "grossly excessive". It also noted that the success in establishing a workable pooling arrangement with the railways had not been helped when, in 1936 and 1937, its effect had been to call for a contribution from the railways to the Board of £10,146 and £48,674 respectively! As with the Thompson Committee, McClintock put this down to the failure of the two parties to agree on the method of co-ordination envisaged by Pole, and to the loss of freight traffic by the Board to traders who opted to carry their own merchandise.

The McClintock committee recommended the creation of a new public authority, consisting of the NIRTB, the NCC and BCDR, and linked to the GNR and SLNCR by a pooling scheme. The stocks of the new authority would be owned by the shareholders of the LMS, the GNR, the BCDR and the Northern Ireland government. It recommended that the capital of the new undertaking should be in the form of equity stock and that it should not be related to the capitalisation of the former undertakings. It should, in fact, be based on their gross traffic receipts which might be taken to reflect "the total amount that the public of Northern Ireland have in the past been willing to pay for transport facilities".

The Belfast Corporation, a profitable concern, should be merged with the new body but only when the latter had met all interest charges and paid the standard dividend on its equity stock for four consecutive years.

However, before any hard decisions on further re-organisation could be made, World War II intervened and the greatly increased traffic put the railways and the NIRTB back into profit and gave everyone a breathing space to consider the matter further.

During the war the government privately considered the idea of selling the NIRTB to the railways, a course of action favoured by them but vigorously resisted by the Board. The government were, in fact, in secret negotiations with the railways

to this end when, for what were probably political reasons, they thought better of the idea and suddenly ended the talks.

Subsequently, the Chairman of the NIRTB and the General Manager of the LMS (NCC), with the full knowledge of the managements of the other railway companies, entered into talks. They proposed a scheme to set up a single transport body for the province to take over the assets of the Board and all the railways operating wholly within the province, and set up a pooling arrangement for those railways, such as the GNR, which operated across the international boundary. Although accepted by the Board and the railway companies, this proposal was not immediately taken up by government which was still wary of the political implications of setting up such an obvious monopoly.

The joint committee of the two houses of the Northern Ireland Parliament reported in 1939. However, its report had little effect other than delaying taking action on the McClintock Report. The only discernible action was to transfer to the government the NIRTB stock held by the railways, in exchange for government securities.

At the end of 1942, a confidential report was submitted to the chairmen of the NIRTB and the NCC by their general managers accepting the principles of the McClintock Report and calling for urgent action to implement them.

The war years were profitable for both the passenger and freight operations of the NIRTB. However, by 1946, freight had moved back into loss, although the passenger side remained profitable.

After the war, all the old problems reappeared for the railways. The first to fall was the BCDR which, because of the configuration of its network, was especially vulnerable to road competition. It also had to meet a huge claim for damages as a result of the Ballymacarrett accident in 1945, which virtually wiped out its reserves, built up from profits made during the war years. The BCDR told the government that, because of its financial position, it would have to shut down its operations completely in 1948.

This finally pushed the government into taking action, which it did by setting up the Ulster Transport Authority in 1948.

Chapter 2

Nationalisation and competition
1945-2000

The Republic of Ireland, 1945-2000

CIÉ 1945-50: a new start

Córas Iompair Éireann was set up by the government of the Republic of Ireland on 1 January 1945. In his speech, moving the second stage of the Transport Boll (1944), Seán Lemass, Minister for Industry and Commerce, described CIÉ as "a new organisation, with a new legislative basis, a new financial structure, a new relationship with the government and the public, and even a new name to signal the break with the old traditions".

Lemass believed that the pre-war position of transport in the country had been unsatisfactory and that, outside Dublin, public transport was inadequate and expensive. The government had been disappointed with the results of the 1932 and 1933 Transport Acts and believed that just changing the GSR structure would not solve the problems. It was decided, therefore, to merge the GSR and the DUTC. Under the Bill, the chairman would be appointed by the state and six directors by the common stockholders. It was thought that this would be a better arrangement than one where the whole board was state appointed.

CIÉ was not to be a nationalised company because Lemass believed that "nationalised transport almost inevitably means subsidised transport". He believed that subsidies were unnecessary and that public transport could be financially viable.

CIÉ's first chairman was Percy Reynolds who had been the last chairman of the GSR. The general manager was E Bredin who had been general manager of the GSR and Frank Lemass, former general manager of the DUTC and brother of the Minister, became assistant general manager.

At its foundation, CIÉ owned over 2,000 miles of railway, about 500 locomotives, 800 carriages, 1,300 wagons, 600 buses, 100 trams, 300 horses and 500 freight vehicles. It also owned the Royal Canal,

six hotels and was the operator of Rosslare harbour. CIÉ was also Ireland's largest employer with a staff of over 21,000.

CIÉ would provide about 80% of the state's public transport services, the rest being supplied by the Grand Canal Company and the various railways whose lines crossed the border with Northern Ireland. Of the remaining 900 independent companies licensed to carry 'for hire or reward', most were one-vehicle concerns and only 28 were bus companies.

Road and rail services improved in 1945. Better supplies of fuel became available as the war ended and, in December, CIÉ announced that branch lines closed in 1942 would be re-opened. This fulfilled a promise the minister had made when the lines were closed. CIÉ would have preferred these lines to have remained closed and replaced by road services but, as sufficient road vehicles were not yet available, they were prepared to see the lines re-opened as a short-term measure. The financial impact of this decision was later blamed for turning a surplus on railway operations in 1945 into a deficit in 1946.

A large scale bus building programme was begun in 1946. Reynolds also announced that the railways would be converted to diesel traction; mainline stopping trains would be withdrawn; branch lines would be closed and replaced by road services. However, it would take CIÉ 25 years to fully implement these strategic proposals.

Although early financial results looked good, by 1948 CIÉ was again in serious financial difficulties. The board recommended to the government that it considered five key measures necessary to remedy the position of the railways:

- Improvement of the permanent way to enable increased speeds to be run with safety.

- Replacement of carriages and wagons and the introduction of diesel traction.
- Closing of branch lines.
- Restriction of private goods vehicles.
- Contribution from the profits of the Omnibus Department.

Reynolds believed that, as the railways were losing valuable freight traffic to the roads, they needed assistance to secure their future. This could take the form of:

- Nationalisation.
- Subsidisation.
- Restriction on other forms of transport.

The Milne Report, 1948

The government decided to deal with the problem by appointing another committee of investigation to be chaired by Sir James Milne, ex-general manager of Britain's Great Western Railway. Milne's brief was:

> to examine and review the position of rail, road and canal transport and report to the minister ... thereon and on the steps it is necessary or desirable to take to secure:
>
> (a) the greatest measure of co-ordination of rail, road and canal transport;
>
> (b) the restoration of the financial position of the public transport companies; and
>
> (c) the most efficient and economical transport system.

Milne's report looked not just at the GSR but at the GNR, the Grand Canal Company and the smaller independent railways. He did not recommend nationalisation or subsidisation, but that the state could give CIÉ relief on liability to repay sums advanced by the state, and by meeting interest charges incurred during capital construction works undertaken in the national interest.

He also recommended the formation of a Central Highways Authority that would be charged with building and maintaining all highways *including the railway's permanent way* and canals (Author's italics).

This Authority would be funded from increased duties on all commercial vehicles and by a levy on rail and canal revenues. If it had been implemented, this radical proposal would have saved CIÉ around £400,000 of the cost of upkeep of the permanent way. However, the government was not interested in the proposal and ignored it.

He also recommended that the Dublin and Stormont governments co-operate so that the UTA and CIÉ jointly acquire the GNR. The reorganised GNR would then take over the other railways operating cross-border services. The first of these recommendations was put into effect in 1950 and resulted in the formation of the GNR Board in 1953, but the second recommendation was ignored. These minor railways – the SLNCR, LLSR and County Donegal Railway – eventually closed. Milne also recommended that CIÉ acquire the Grand Canal Company.

Milne supported maintaining the railways and was against closure of branch lines. Forty-three branch lines had been targeted for closure by CIÉ. However, he regarded them as part of the national system of highways and believed they should receive the same consideration as ordinary roadways.

Milne did not believe that diesel-electric locomotives were appropriate for Ireland. As we have seen, in 1947 CIÉ had announced its intention to convert to diesel-electric traction, but Milne regarded this as unproven technology (in spite of its extensive use in the USA and mainland Europe). Therefore, he recommended that CIÉ build 50 new steam locomotives. For smaller trains, it should buy a limited number of diesel railcars like those being bought by the GNR. He also disagreed with the plan for a new central bus station in Dublin and for the plans to build, with Leyland's help, a chassis building facility.

Essentially, Milne fundamentally disagreed with the strategy that CIÉ had put forward to provide a modern transport service. CIÉ's management, in their turn, were dismissive of the Milne report and stuck to their view that there was not enough traffic to allow the railways to survive if unrestricted road transport was permitted.

Under state control, 1950

In February 1949 Percy Reynolds was removed as chairman of CIÉ (Milne had criticised his over-powerful position as chairman). The government also announced the immediate acquisition of the transport system, road, rail and canal (other than the cross-border lines) even though this was not one of Milne's

Line up of GNR AEC railcars at Great Victoria Street station, Belfast.   Author's collection

recommendations.

Reynolds was replaced by TC Courtney. Courtney appointed GB Howden, general manager of the GNR, to the post of general manager. Howden would later go on to become chairman of the UTA; OVS Bullied, latterly Chief Mechanical Engineer of the Southern Railway in England, was appointed CME.

Work on the Broadstone bus body-building shops, the Inchicore chassis-building factory and the Store Street Central Bus Station had been suspended while the Milne report was awaited. Although a contract with Metropolitan Vickers for six diesel-electric locomotives had been cancelled, Sulzer had delivered the engines. These lay in Inchicore for several years before CIÉ decided to ignore Milne and go ahead and dieselise anyway.

The 1949 Transport Bill became the 1950 Transport Act and CIÉ Mark 2 was born. The government could now appoint the entire board. CIÉ was released from all statutory controls over fares and rates. The final decision on branch line closures would be left to CIÉ, subject to the approval of a Transport Tribunal set up under the Act.

The new CIÉ began to invest and innovate. Together with the GNR, they introduced a Belfast-Dublin-Cork 'Enterprise' express train. They ordered 60 AEC diesel railcars similar to those already acquired by the GNR. They also ordered four Walkers diesel railcars, similar to those in use on the County Donegal Railway (P118), to modernise the West Clare narrow gauge line. The use of the 17-year old Drumm battery trains was ended. New coaches and freight rolling stock were constructed. Fifty double-deck buses, 159 single-deck buses and 20 coaches were ordered to replace obsolete vehicles and allow the expansion of services. The Central Bus Station project was given the go-ahead.

In 1951 it was also announced that the GNR was to be acquired jointly by the Dublin and Stormont governments, although it was two more years before this deal was finalised.

In 1956 an experiment was begun to see if certain branch lines could be made viable using lightweight diesel locomotives. Three Deutz 'G-class' diesel locomotives were bought and put into service on the Newmarket, Gortalea to Castleisland and Clara to Banagher branches during 1956-57.

In contrast to railway operations, CIÉ's bus operations were profitable. The bus fleet grew from 1,014 vehicles in 1951 to 1,153 in 1958. There was

CIÉ pioneer Sulzer locomotive No B113 on display at the 'Inchcore 150' celebrations.   Author

CIÉ Deutz locomotive G601 on display at 'Inchicore 150'.   Author

substantial new bus building during 1953-54 and, in 1953, the new Central Bus Station, or *Busaras*, at Store Street was opened.

Therefore the 1951-57 period saw the start of the conversion of railways to diesel traction using both diesel railcars and 94 Metropolitan Vickers diesel-electric locomotives, development of road services, co-operation with the Northern Ireland government in the operation of the ailing GNR through a joint Board, some closures of branch lines and the absorption of the Grand Canal Company. However, it did not see any reduction in CIÉs financial problems.

The Beddy Report, 1956

By 1956 it was clear that the 1950 Act had failed and a new committee was formed under Dr JP Beddy to examine the whole transport situation again.

Beddy looked at the arguments for completely closing the railway system, but concluded that there was traffic suitable for the railways. It would be better to reduce the network to a size that would be able to economically handle this level of demand. He did not feel that widespread road congestion would follow from the transfer of traffic from rail to road. Neither did he agree that private transport should be restricted.

Having rejected abandonment of the railway system, subsidisation and forcing traffic on to the railways, Beddy stressed that the emphasis should be placed on making the maximum use of the special advantages of each mode of transport. Traffic should not be carried by one form of transport that could be more economically carried by another. He blamed CIÉ for losing traffic to private carriers by not developing vigorous, enterprising, and efficient road transport services. He toyed with the idea of recommending the establishment of a separate state-owned road transport body, but then decided against it.

When published in 1957, the report recom-mended that the length of railway and number of stations should be reduced in such a way as to ensure that as many as possible of the areas served by CIÉ should be within a short motor (public or private) journey from a station.

The report indicated that the new network would probably consist chiefly of the main lines. It included a map showing a reduction in railway mileage from 1,918 miles to 850 miles and a 75% reduction in stations. In fact, the present (2000) railway network is very close to that envisaged in the Beddy report. Rather than indicating which lines should be closed, Beddy indicated which should, in his opinion, be kept open. He recommended that the Transport Tribunal be abolished and that CIÉ be given the power to close lines. A Transport Council should be set up, but this would only have an advisory role.

Beddy believed that, if his recommendations were implemented, greater operating efficiencies would follow for the railways. This would allow greater economies in operation and a focus on types of traffic for which the railways were best suited. This in turn would allow the railways to operate without incurring serious financial losses, so weakening the whole structure of public transport.

Beddy also attacked the restrictions imposed on CIÉ by its 'common carrier' obligations. Although these had been modified in the 1950 Act, they were still largely in place. Beddy recommended that CIÉ should be freed from these restrictions and be able to quote whatever fares or rates it considered appropriate.

In the debate which followed the publication of the report in 1957, CIÉ management made it clear that they were still not of the opinion that shrinking the railway system was the solution, but rather that traffic should be forced off the roads and on to the railways.

The Minister of Transport, Seán Lemass, commented that unlike what was happening in Northern Ireland, the Dublin government would not be making the final decision about the size of the railway system. He proposed a new Transport Bill that would give CIÉ the powers to close lines or stations without having to obtain the approval of a Transport Tribunal, as Beddy had recommended.

The 1958 Transport Act

Two Transport Acts were passed in 1958, one to give effect to the Beddy proposals, the other to merge the southern portion of the GNRB with CIÉ. The 1958 Transport Act in a sense created CIÉ Mark 3.

The main changes resulting from the passage of the two Bills into law were to:

• Merge the GNR with CIÉ.

• Restructure CIÉ's capital to improve its financial position.

- Remove the 'common carrier' restrictions.
- Provide CIÉ with an annual grant of £1,175 million for five years.

No changes were proposed in the size of the railway network; that would be left to CIÉ. The company was given until 1964 to bring expenditure into line with revenue. No new restrictions were placed on road freight operators.

TC Courtney was replaced as chairman by Dr CS Andrews, previously managing director of Bord na Mona. His first task was to handle the absorption of the GNRB into CIÉ. This resulted in CIÉ acquiring 2,500 staff, 120 miles of railway, 83 steam locomotives, 24 railcars, 119 passenger carriages, 247 other items of rolling stock, 160 buses, 300 road freight vehicles and 11 Howth trams. The balance of the GNR fleet of railway locomotives, railcars and other rolling stock was transferred to the UTA.

As part of a major management overhaul, Andrews divided CIÉ operations into a new 'area' structure each headed up by a manager with the responsibility for dealing with problems at a local level. He also established a central Transport Control and Planning Office to control the integration of road and rail services.

He introduced a more flexible approach to pricing which in today's business environment would be taken for granted. Andrews recognised that in a very competitive environment, the customer had to be attracted by the advantages of the service being offered. It was no longer possible to force the customer to use the service. He placed great emphasis on encouraging CIÉ to negotiate contracts for the handling of a customer's entire transport needs. Although, in public, Andrews expressed his support for the railways, in practice he was prepared to close uneconomic lines. Under his chairmanship, CIÉ began to seriously prune its rail network.

In 1959 it was decided to close the Sallins-Tullow branch line and the Harcourt Street-Shanganagh Junction line. Although heavily criticised today as being short sighted, given the way the Dublin suburbs have expanded along the route of the Harcourt Street line, in 1959 the closure did not seem to attract much attention. The abandonment of a number of GNR 'stump' lines in Cavan and Monaghan, along with the little-used Headford Junction-Kenmare and Farranfore-Valentia Harbour lines in Kerry, were also authorised. These closures caused little controversy.

In 1959 CIÉ began an investigation of uneconomic railway routes and services. Before a closure was made, CIÉ would give assurances that there would be satisfactory replacement road services. This was not a legal requirement imposed on CIÉ, but it believed that, as the national carrier, there was an obligation to provide these services.

In 1960 CIÉ decided to close the Macmine Junction-Muine Bheag/Waterford line. As it turned out, this line was not closed until 1963. It also decided to close the Waterford-Tramore line, the whole of the West Clare narrow gauge system, and the Cork, Bandon and South Coast line. There were many objections to these latter closures, particularly of the Cork lines where a local businessman even took legal action against CIÉ to have the lines retained. The Tramore line closed in December 1960, the West Clare lines in January 1961 and the Cork lines in March 1961. Even to this day, there is much resentment in Cork over both the fact of closure and the way that local objections were ignored.

In 1963 the next batch of closures took place. These were the Portlaoise to Kilkenny, Portlaoise to Mountmellick, Castlecomer Junction to Deerpark, Kilfree to Ballaghderreen, New Ross to Macmine Junction, Muine Bheag to Palace East, Clonsilla to Navan, Navan to Oldcastle, Athy to Ballylinan, and Enfield to Edenderry lines. Passenger services were also withdrawn on the Cobh Junction to Youghal, Limerick to Tralee, Ballingrane to Foynes, and Waterford toNew Ross lines. These closures did not provoke much opposition. In addition to the closures of lines, many stations on the network were also pruned during the early 1960s.

By 1964 it had been calculated that the 29% reduction in the railway network had resulted in only a 5% fall in revenue.

The next closures were in 1967 with the Thurles to Clonmel, Rath Luirc to Patrickswell and Mallow to Waterford lines. At the end of this process the railway route mileage had been reduced from its 1958 figure of 2,700 miles to 785 miles.

It would be a mistake to see these closures as a running down of the rail network, as a prelude to a possible total abandonment, as was happening in Northern Ireland. In 1963 the Minister of Transport and Power, Erskine Childers, announced the

Preserved CIÉ Metropolitan Vickers 'A' class locomotive A39 in original silver at 'Inchicore 150'.   Author

government's intention to maintain the main arterial railway system of the country. The period 1960-62 saw the introduction of 52 General Motors B-class diesel locomotives, 14 additional Maybach E-class and seven additional Deutz G-class shunters. From 1 April 1963 all steam traction was withdrawn. In addition, there was significant investment in new rolling stock, both passenger and freight, modernisation of station buildings, track and equipment, and significant new passenger marketing initiatives were begun.

As part of the new 'modern' image, in 1961 railway passenger rolling stock livery was changed from green to orange and black. Bus livery was altered from green to dark blue and cream for double-deckers and red and white for single-deckers. In 1964 the CIÉ 'Flying Snail' emblem was replaced with the 'Broken Wheel' logo.

Road passenger services

Road passenger services experienced much less upheaval during these years than did the railways. Buses were built on Leyland chassis at the Spa Road works. Between 1948, the year the contract for the

supply of chassis by Leyland to CIÉ had been signed, and 1964, Leyland had supplied CIÉ with 600 single-deck and 750 double-deck chassis.

In 1960-61 CIÉ had introduced a fleet of 60 74-seater double-deckers and in the following financial year 58 single-deckers. They also adapted 53 buses for one-man operation. Major industrial relations problems followed from both the introduction of larger buses and one-man buses.

In 1963 CIÉ tried to introduce one-man single-decker buses onto 100 provincial routes. The unions objected strongly and demanded major concessions. CIÉ refused and there was a strike. The government asked Andrews to postpone the introduction of one-man buses, which he did. It would be four more years before CIÉ would introduce one-man buses in a limited way on some provincial services.

In 1961 CIÉ introduced a new type of express bus service. Coaches would be used and passengers would buy their tickets before boarding. The unions agreed to driver-only operation because it was a new type of service and, if successful, CIÉ promised that the concept would be extended to other parts of the country.

Line of ex-CIÉ and ex-GNR steam engines on Dundalk scrap line in 1959.   David Lawrence

The initial service was from Dublin to Cavan, a route which had been part of the GNR bus and rail services. It was a success, cutting an hour off the stage carriage schedule. Over the next couple of years, driver-only express services were operating to Enniskillen, Derry and Letterkenny, all towns that had recently lost all or part of their rail connections. These services were the foundation of what later became the 'Expressway' division of CIÉ /Bus Éireann.

The Pacemaker Report, 1963

In 1962 CIÉ began a major study of the options for public transport. This report, which would become known as the 'Pacemaker Report', was published in 1963. Its main elements were as follows:

- None of the alternative public transport systems available were so superior to the others as to allow immediate action to be taken (switching from one to the other) with short term favourable results.
- Given the environment in which it was operating, CIÉ faced the problem of either providing its service at the expense of solvency, or of being solvent by refusing to provide a service.

- Private transport in urban areas was becoming self-strangulating.
- Public transport policies were a compromise between what was theoretically satisfactory and what was acceptable in practice. Thus, it rested with national transport policy-makers to state the degrees of compromise required.

The report also analysed the profitability of CIÉ's transport network. The findings were as follows:

- Rail: 410 miles of track were wholly profitable, 300 miles were marginally profitable and 750 miles unprofitable.
- Dublin City Bus: Overall profitable, although considerable cross-subsidisation was taking place. Of the routes, 36 were profitable, eight marginally profitable and 35 loss-making.
- Provincial Bus: Only 72 out of 234 services wholly profitable.

For the future, Pacemaker identified two main alternatives:

- An all road system.

- A combined road/rail system where mainline rail would carry freight only. All passenger carriage would be by road.

The combined road/rail alternative was seen as an interim stage to an all road system.

However, the idea of closing the railway network completely was politically unthinkable, especially as the minister, Erskine Childers, had already stated his belief in the preservation of an arterial rail network. He had also indicated that the Dublin commuter rail system would probably remain essential even if it had to be run at a loss. This was a sign that, even in the early 1960s, Dublin traffic congestion was becoming a serious problem. Childers also stated that public transport could not be governed merely by the laws of supply and demand and that the government must retain the right to determine the overall direction of transport policy.

In 1964 a Transport Act was passed which, in spite of Pacemaker, made little change to the operating environment of CIÉ. It was now clear that the state wanted the existing rail network to be substantially retained. It also accepted that railways could not be run profitably and that profits from road services could only partly offset rail losses.

The main value of the Pacemaker report, therefore, was to force the government to face a reality that it had previously tried to ignore and to lift the burden of trying to achieve the financially impossible from CIÉ's management.

In 1966 Andrews retired from his post of chairman and chief executive and was replaced as chairman by TP Hogan, who had been a non-executive member of the CIÉ board since 1957. Seán Lemass, who had been general manager for the previous 20 years, became chief executive. In 1966 CIÉ received 12 additional General Motors diesel locomotives – the 181-class.

The population of Dublin had doubled since the 1930s and the number of private cars had trebled since the 1950s so, by the late 1960s, traffic in Dublin had become a major problem. Bus schedules were being severely disrupted, particularly at peak times.

Attempts were made to deal with the problem by introducing one-way systems, parking restrictions and later traffic wardens. CIÉ introduced radio control to its Dublin bus fleet from the mid 1960s. However, this did not provide a solution.

In 1967, as the result of the findings of a CIÉ Origin and Destination Survey, Erskine Childers was able to state that basically CIÉ's Dublin bus services were properly structured. Since the pattern of services was dictated by the road structure, no dramatic change could take place until, and unless, Dublin Corporation adopted proposals for ring-roads, through-roads or new bridges.

New Leyland Atlantean double-deckers were introduced in late 1966. From 1966-1974, over 600 Atlanteans were put into service by CIÉ. This choice of bus, on top of Dublin's traffic problems, led to many operational headaches for Dublin bus service managers. The Atlantean design was seriously flawed and the Atlantean fleet gave endless mechanical problems during its life. In many ways the Atlantean design, and Leyland's attitude to customer service during the 1970s, marked the beginning of the end of Leyland as the dominant bus supplier in the UK and Ireland.

During 1967-68, the government and CIÉ introduced a free school transport scheme. This required CIÉ to quickly find 300 buses from wherever they could, the balance of 495 being privately owned vehicles. In 1968 CIÉ put in place a programme to build 240 specially designed school buses.

The First McKinsey Report, 1970

In 1970, as the result of CIÉ's growing financial deficit, the government commissioned a new report from the management consultants McKinsey and Co.

The McKinsey report was published in 1971 and stated that CIÉ's losses were due to a number of factors. The value of the government's subvention had been eroded by inflation, costs had increased but government had blocked increases in rates and fares. Competition from road transport had also increased. This had resulted in a fall in the number of passengers carried by Dublin bus services and a consequent fall in revenue. There was also a rapid increase in the losses incurred by the railway. As those two services accounted for over 60% of CIÉ's revenue, these trends had had a major impact on overall results.

McKinsey carried out three analyses to define a new strategy for the railway. These were as follows.

Firstly, McKinsey examined the railway's financial losses. It was concluded that, in spite of

CIÉ Atlantean double-deckers. These mechanically unreliable buses gave CIÉ severe problems.   Photobus

large-scale modernisation and system rationalisation, losses had increased rapidly and had reached a level which cast doubts on the long term viability of the railway. Although the amount of freight, and number of passengers carried, had increased between 1965 and 1970, this had not resulted in reductions in operating losses. In addition, the commuter traffic was completely loss-making and, if looked at from a purely commercial point of view, would have been ended.

Secondly, McKinsey conducted an identification of the social benefits of the rail system. Once social benefits were included, the conclusions from the purely financial analyses were altered. A cost-benefit analysis of Dublin rail commuter services showed that they provided major social benefits, reducing road congestion and travel time for both rail and road commuters. The higher costs which would result from increases in journey times, following on from the withdrawal of rail commuter services, would have more than outweighed any financial benefits.

McKinsey concluded that the total economic costs of all other passenger services outweighed their economic benefits, but only by a fine margin. There were also social benefits in carrying freight by rail

which to some extent offset the higher cost of railway freight services over road services. However, even taking into account the social benefits of rail services, the total economic cost to the country of keeping the rail system exceeded the social benefits produced.

The third analysis was a calculation of the transitional costs incurred if the railways were closed. McKinsey looked at two options. One was to close the whole railway network immediately, and the second to close it over a five year period. On purely financial criteria it was concluded that the costs to the nation of closing the railway completely would be greater than improving financial performance by investing in improvements to the rail system.

As a result of these analyses McKinsey recommended streamlining railway operations to bring economic costs and social benefits more into line with each other by reducing operating costs and boosting revenues. Several major changes would be needed to achieve this result:

- Close many small freight and passenger stations and freight-only lines. This would have little adverse effect on revenues but allow major economies to be made in both train and station working costs.

- Lightly loaded stopping passenger train services should be withdrawn and substituted with bus services.
- Routing of some passenger services should be rationalised to eliminate route duplication.
- Upgrade Dublin rail commuter services by the provision of new rolling stock. This would contribute towards the reduction of the city's traffic congestion problems.
- Invest to rationalise and develop freight services through the provision of new rolling stock and mechanised handling equipment.
- A phased reduction in the numbers of employees, preferably through normal wastage. As staff accounted for 70% of total costs, if this was not done, the current level of losses would not be reduced.
- CIÉ's road freight division should be allowed to compete with rail freight to avoid traffic being lost by CIÉ to private hauliers.

It was recommended that the railway should be given an annual grant to cover the costs of track maintenance and renewal. This would recognise the contribution made by the railway to the total economy. The current system of financing should be replaced by one in which above-the-line grants would be paid to finance socially beneficial but economically unviable activities. These would include much of the rail network and many rail commuter and bus services.

These payments would allow CIÉ to make a small working profit before charges. Provided CIÉ was allowed the freedom to set rates and charges, McKinsey believed that the recommended changes would eventually allow the railway to pay its financial charges, including charges on new investment, and still make a small profit.

In order to ensure that the most effective use was made of bus and rail services in solving Dublin's transport problems, it was recommended that they should be run as a single operation with combined scheduling, pricing and marketing.

However, the government, with many other problems on its plate, postponed implementing the recommendations of the McKinsey Report.

In 1971 attempts were made to improve traffic flows in Dublin by introducing an experimental bus lane, putting on extra buses and suburban trains, and offering reduced fares on some Dublin bus commuter routes. Although the results were positive from CIÉ's point of view, the Gardai were not impressed by the bus lane experiment and was be ten years before it was tried again.

Between 1968 and 1972, the original Metropolitan Vickers A and C-class locomotives, whose diesel engines had turned out to be very poor, were re-engined with General Motors power units.

Dublin's suburban rail traffic was also growing and CIÉ carried out alterations at Connolly and Pearse stations, to permit easier through workings. Questions began to be asked in the Dail about the possibility of expanding these services.

In 1972 a new general manager, John Byrne, was appointed to succeed Daniel Herlihy. Herlihy had taken over from Frank Lemass who had retired in 1970.

Railplan 1980 (1972)

After the publication of McKinsey, CIÉ drew up 'Railplan 1980', a study into how the deterioration in the financial position of the railways could be reversed. The study was completed in late 1972 and set out a way by which CIÉ could deliver a modern, efficient and more economic railway system by the year 1980. It stated that:

- There should be a major reduction in railway staffing.
- The number of passenger and freight handling stations should be severely cut.
- Remaining freight-handling facilities should be mechanised to a much greater extent.
- The concept of the 'block train', a standard freight or passenger train which would travel unchanged from origin to destination, should be adopted. This would reduce the need, and consequent delays associated with shunting and would allow better utilisation of rolling stock and locomotives.
- Push-pull configurations should be used for suburban workings. This would reduce the turnaround time for trains at termini by eliminating the need for locomotives to 'run round'.

Plans to modernise and possibly electrify the Dublin suburban rail network were under consideration by a different group at this time.

Early in 1973, CIÉ began to re-organise its rail services. Heuston station had two new platforms added and trains to the west now began there rather than from Pearse, which was able to specialise in the growing Dublin commuter traffic. Belfast and Sligo trains would leave from Connolly.

Between late 1972 and early 1973, 71 new air-conditioned coaches arrived from British Rail Engineering Ltd. These were used on the new 'Supertrain' services introduced in the summer of 1973. These Supertrains were lighter and shorter than previous train configurations, and allowed a faster and more frequent service to be provided. The number of trains on the Dublin-Cork, Belfast, Limerick, Rosslare and Galway services was increased. At the end of 1973, CIÉ was able to report an increase of traffic of 8% over the previous year's figures, which in turn had been 6.5% up on the 1970-71 figures.

Older stock was cascaded to use on commuter services. Some of the AEC railcars introduced in the 1950s had their engines removed and, using a locomotive, formed push-pull configurations on suburban services. The number of cross-town, peak period services was also increased. Closed stations – Sidney Parade and Booterstown – were re-opened, and a new station, Bayside, was built, all between 1973 and 1975. Automatic ticket checking was also introduced at some commuter stations. At the end of 1973, CIÉ was able to report that Dublin suburban train service loadings were up by 7.5% over the previous year.

The 1973 annual report also noted an increase in Dublin bus service loadings of 4.5%, in spite of deteriorating traffic conditions in the city.

In 1973 a Dublin Transportation Study was published and the government accepted its recommendations in principle in 1974. For the railways it recommended:

- Services on existing routes should be improved by increasing peak hour capacity.
- Parking provision should be made at suburban stations.
- Feeder buses to suburban stations should be provided.
- Suburban services with a number of new halts should be provided on the Galway and Cork lines as far as Blanchardstown and Clondalkin.

A CIÉ push-pull train at Rush and Lusk on 7 April 1984, with a Dublin to Drogheda outer sunurban working. The train is hauled by rebuilt Metro-Vick locomotive No 218.

Michael McMahon

At the end of 1973 CIÉ, along with the Department of Local Government, Dublin Corporation and An Foras Forbartha (the Institute of Development), commissioned a study to look into the feasibility of setting up a Rapid Rail Transport System for the capital. A group of internationally-renowned consultants were appointed to examine the problem and report within a year.

In 1973 TP Hogan retired as chairman and was replaced on 1 January 1974 by Liam St John Devlin who moved from the chairmanship of the board of B and I, the state shipping company. Devlin thought that CIÉ needed to rethink its whole role as a public transport provider. He believed that the company was seriously over-manned and that staff reductions greater than those envisaged in Railplan 1980 were required. He also committed himself to a total system of one-person-operated (OPO, OMO no longer being politically correct) buses.

In 1974 the Minister for Transport and Power reported that the government had decided that the railway system should be preserved, subject to further concentration and re-organisation and announced 'approval in principle' for a £23 million programme to modernise the railway in the light of the McKinsey and Railplan 1980 recommendations.

In 1975 the minister stated that, in his opinion CIÉ's main business should be in running buses and trains and that they should concentrate on these and cut back on peripheral activities. Devlin agreed with this view.

As a result, by the end of 1977, CIÉ had disposed of its interests in Shannon cruisers, Irish Ferryways, Irish Continental Line and Aerlód, its air freight subsidiary. In 1984, after a number of years of poor financial results, CIÉ disposed of its hotel interests. CIÉ would have liked to dispose of its responsibility for running Rosslare harbour but, because of the complexity of the legal situation regarding its ownership and control, it was decided that such disposal would be more trouble than it was worth, so nothing was done. Rosslare harbour remains a profitable part of CIÉ.

Four more branch lines were closed in 1975 – Ardee to Dromin Junction, Collooney to Claremorris, Loughrea to Attymon Junction and Listowel to Ballingrane. The Loughrea branch had carried passengers, the other three freight only.

In 1976 CIÉ ordered 18 new locomotives, the 071-class, from General Motors. These powerful new

CIÉ 071 class locomotive No 072, displaying the first Irish Rail logo.   Author

locomotives were introduced into service in 1977, after some delay due to industrial action.

As the emphasis in the Railway Development Plan was on the establishment of a radial network of lines based on Dublin, it was decided in 1976 that passenger services on the Limerick-Claremorris line would cease and be replaced by buses. Freight and Knock pilgrim traffic, however, would be retained. An attempt to close the Rosslare Harbour-Waterford-Limerick Junction line ran into heavy local and political opposition and had to be abandoned.

In 1977 Devlin summarised CIÉ's three-strand approach to upgrading the radial network as:

- A substantial improvement in track to allow faster running. The track renewal programme would have to be a major one since there had been no significant investment in the permanent way since the 1930s.
- The automation of signalling to increase track capacity. The installation of a new signalling system, called CTC (Centralised Traffic Control) based at Connolly station, was begun in 1974 and the first section, from Inchicore to Ballybrophy, came into use in 1975.
- The provision of modern coaching stock on all routes to improve comfort levels and attract more custom.

The target of renewing coaching stock was proving difficult to meet because of government delays in approving CIÉ's plans regarding how best this should be done. CIÉ had wanted to enter into an arrangement with the German company Linke, Hoffman and Busch under which they would take over a large part of the Inchicore railway works complex and establish a carriage manufacturing plant there. This would supply both CIÉ's needs and provide LHB with an Irish plant from which they could supply customers further afield. It was a similar scheme to the one entered into with Van Hool-McArdle for a bus-building project at CIÉ's Spa Road site. A proposal based on the LHB scheme was placed before government in 1977.

By 1978 the coaching stock problem was becoming acute and Devlin met with the minister to discuss:

- Purchase of new mainline coaching stock.
- Permanent way renewal and electrification of the Dublin suburban lines.
- An indication of the government's future policy towards the railways.
- The future role of public transport in the conurbations.

In 1977 CIÉ had unveiled its proposals for the electrification of the Howth-Bray line. This would be the first stage in the establishment of a rail rapid transit system for Dublin. However, approval from government was not given until May 1979. The delay in approving new coaching stock, coupled with increasing deficits incurred by the railway side, led some senior people in CIÉ to suspect that the government's commitment to the preservation of the railway network was cooling and that a policy of attrition was being used to gradually kill it off. The approval given for the Howth-Bray electrification project allayed these fears to some degree.

Bus operations

On the bus operation side, CIÉ in the 1970s operated over 2,500 buses carrying over 300 million passengers a year. Dublin bus services in the 1970s were severely affected by both increasing traffic congestion in the capital and severe industrial relations problems. Provincial bus services experienced fewer congestion and labour problems and were a success story for CIÉ. Passenger numbers doubled in the decade to 1975 from just under 20 million to almost 47 million. The number of private cars had grown by 70% over the same period.

In the cities, the big problem was with traffic congestion. Dublin's population had increased from 470,000 in 1900 to 1,350,000 in 1991. Between 1955 and 1979 the number of cars in Dublin increased six-fold. In Dublin, traffic management devices had been applied to only a limited extent, and the option of forcing a restriction on the number of cars using the city had not been used at all.

Dublin was not unique in this respect in the British Isles. However, in Ireland the political problems of confronting citizens with severe restrictions on the use of their private cars were too daunting, for politicians seeking re-election under a voting system where a few hundred votes could mean the difference between keeping or losing one's seat.

In the fashion of the times, new urban satellite

developments, built in the 1970s, were served by roads designed to facilitate private transport, rather than by public transport. In the Tallaght development there was a proposal for a dedicated busway, and land was even set aside for it. However, it was never built. In 1976 CIÉ reported that traffic speeds in Dublin were amongst the lowest in Europe, about 6 mph during peak hours.

Dublin began to install radio control for buses in 1969, having used a telephone-based system for inspectors from the 1950s in an attempt to deal with the problems of traffic congestion. They followed this in the mid 1970s with the development of a sophisticated Automatic Vehicle Monitoring system (AVM), a computer-based system which covered the whole fleet. This allowed the position of all the buses to be displayed at a central control room and then corrective action could be communicated by radio to the drivers. However, because of resistance from bus crews, it was 1983 before the entire Dublin city service was under AVM control.

Attempts to introduce OPO double-decker buses continued to be resisted by the unions during the 1970s. In fact, industrial relations problems were a continual source of frustration for CIÉ management in the operation of its Dublin city services, with bus crews routinely resisting operational and technological changes, with consequential financial losses for the company.

In the 1970s there were calls for the establishment of a traffic management authority for the city, but no action was taken on this proposal until the 1980s.

In 1977 the process of establishing regulations to allow the introduction of bus priority schemes was begun, but it was 1980 before the first short contra-flow bus lane was introduced in central Dublin. Further bus priority measures would be introduced during the 1980s and, by 1984, 67 bus priority schemes had been introduced totalling about ten miles. In addition, the government accepted a recommendation from the Transport Consultative Commission's report in 1980 for the establishment of a Dublin Transport Authority, but it would be another six years before this was set up.

In 1975 the success of the OPO express bus service, begun in 1961, was enhanced by the introduction of the 'Expressway' brand name. The growth of this provision over subsequent years was in the order of 25%-30% per annum.

In some cases, express bus services did parallel rail routes, but the appearance of direct competition was misleading for two main reasons. Firstly, mainline trains were operating with fewer intermediate stops and thus some places were finding that the fall in their rail services was being in part compensated by the provision of the express bus service. Secondly, Expressway was targeting discrete market sectors, such as the youth market, where demand was much more price-sensitive and much less time and comfort-sensitive than were the main rail market sectors. Thus, long distance bus services were designed to complement rather than compete with the railway provision.

Where competition did exist between bus and rail, it was with private operator provision, which also competed with CIÉ's express bus services. On occasions when CIÉ had tried to provide bus feeder services to the railways, private operators would provide competing direct services from point to point, usually to and from Dublin.

In 1975 a May to September, weekend, Galway-London service was commenced in co-operation with the National Bus Company in Britain. It served Athlone, Dublin, Birmingham, Coventry and Watford. Another initiative was the establishment of the 'Interlink' service where express services were linked at various transfer points to allow cross-radial journeys to be made. Cross-channel and Interlink services were also combined under the 'Supabus' brand name to allow a greater choice of origins and destinations. Both Supabus and Interlink services were very successful.

The Van Hool-McArdle project, 1973-77

During the 1970s CIÉ had serious mechanical problems with its Leyland bus fleet. As was noted earlier, CIÉ had entered into an agreement in 1947 which bound them to Leyland for the provision of all road passenger vehicles which fell within the Leyland range. In return, Leyland agreed to supply CIÉ with spare parts at cost price and help CIÉ to set up a chassis-building factory. In the event, the plans for building this factory were dropped. This agreement, which did bring benefits to CIÉ, was renewed several times until 1974.

CIÉ was happy with the Leyland equipment they

Leyland Atlantean with Van Hool-McArdle bodywork. A total of 238 of these buses entered service between 1974 and 1979.
Photobus

CIÉ KE class Bombardier single-deck bus in 'Expressway' colours. These 45-seat buses entered service in 1981.
Photobus

had bought up until 1964, but after that date they noted a serious deterioration in quality. C-class and M-class single-deck buses suffered severe mechanical problems, but the worst affected was the 600-strong Leyland Atlantean double-decker fleet. Leyland equipment failures cost Dublin city services dearly for about ten years.

As a result of these problems, CIÉ decided that a Belgian company, Van Hool, would take over the Spa Road bus-building works in 1973 and that the Leyland contract would be ended in 1974.

In line with Devlin's attitude mentioned above, it was no longer believed that bus building was an activity suitable to a transport operator. As CIÉ's demand for buses was not enough to keep the Spa Road works in full production, Van Hool hoped to develop export orders for its Irish plant. Thus, in 1973 Van Hool teamed up with the Dundalk firm of Thomas McArdle Ltd and set up Van Hool McArdle. They leased the Inchicore body-building workshops and took over 248 CIÉ staff.

The first Van Hool McArdle buses were based on the Atlantean chassis and engine. This was regarded as an improvement on the Leyland product. However, CIÉ was not satisfied with Leyland, as the Dublin city fleet had experienced as much as a 25% failure rate in fleet availability. In co-operation with Van Hool, CIÉ set about finding more reliable equipment and eventually settled on a combination of General Motors engines and Allison transmissions, with a combination of Cummins engines and Voith transmissions as a back up. As a result, they cancelled the Leyland contract for power units. It was intended that Van Hool would build all CIÉ's buses for a period of five years.

CIÉ planned a 'family' of buses – double-deck, city single-deck, and intercity single-deck – which would be specially designed to meet all their major fleet needs. They would prepare the preliminary designs and specifications and Van Hool would build the buses based on these designs.

Van Hool soon found the costs of the Spa Road facility too high and production capacity too low, and decided to build a new plant with a 600 buses per year capacity. Van Hool believed that the new plant and bus designs would increase export opportunities. CIÉ were happy with this plan as they wanted to move away from the Leyland design to new types as

quickly as possible.

However, by 1976 the agreement between CIÉ and Van Hool was in trouble. The problem was the initial 'cost-plus' contract between the two parties. CIÉ found that it was impossible to predict the final cost of a bus from a 'cost-plus' contract, nor was there any incentive for the manufacturer to maximise efficiencies. CIÉ were no longer prepared to buy off a 'cost-plus' contract and wanted to negotiate a new fixed-price contract, with cost escalation clauses, for the proposed new bus 'family'. Van Hool would not sign the fixed cost contract and CIÉ suspended its agreement with them. The legal battle over this cancellation rumbled on until 1992 before being finally settled in CIÉ's favour.

The Bombardier project, 1978-85

Following a report on Dublin's traffic, the government authorised expenditure of £40 million on renewing CIÉ's buses. In 1977 CIÉ asked a bus designer, Otto Schultz of Hamburg Consult, to design a new bus 'family' based on GM engines, Allison transmissions and Rockwell axles. They negotiated a contract with American Motors General, a subsidiary of American Motors, to build the buses. American Motors leased a factory at Shannon but, at the last moment, AMG pulled out of the deal and CIÉ asked General Motors to take over. GM had been approached in 1973 but had not wanted to be involved in building buses that were not to their own design. In 1977 this remained their position.

An executive of American Motors then suggested a joint venture between Bombardier, a Canadian bus builder, and the General Automotive Corporation. Thus was created Bombardier (Ireland) who took over the Shannon plant and with whom CIÉ signed a contract.

In 1981 delivery of the new Bombardier buses began. By December 1982 there were 52 express single-deckers and 276 double-deckers in service. Eventually 365 double-deckers were built. Between 1983 and 1985, 202 high capacity city single-deckers were built and, between 1983 and 1987, 227 rural single-deckers went into service. The Bombardiers allowed a 50% increase in average mileage per bus and a similar percentage decrease in maintenance costs per mile. The number of maintenance staff was thus reduced.

Bombardier-built bus at Broadstone depot, Dublin, 366 of which were built for CIÉ.   Photobus

The initial optimism with the Bombardier fleet would later fade as basic problems with the design emerged and as a result of a dispute involving the company, CIÉ and the government. Structural problems emerged after vehicles had been in service for about two years and these proved expensive to sort out.

The dispute between Bombardier, CIÉ and the government arose because of production problems. CIÉ required only 150 buses per annum, but the Shannon factory could only be operated economically at an output of 250 buses a year. Bombardier had hoped that the surplus output would be absorbed by export orders, but these proved hard to obtain and eventually CIÉ agreed to accept more than 150 buses each year. Bombardier understood that CIÉ would accept 250 buses in 1982. However, the government authorised a capital expenditure programme which would not allow the purchase of such a number. As a result, the numbers employed at the Shannon plant had to be cut.

In 1983 Bombardier decided to sell its stake in the Shannon plant to GAC. GAC withdrew in 1985 and CIÉ operated the plant until 1986 when it was closed down.

After this second failure in bus building arrangements, CIÉ changed its policy to one of inviting tenders and buying standard models on the basis of price and quality.

The Foster Report and the Second McKinsey Report, 1979

In 1978 the National Economic and Social Council (NESC) asked Professor Christopher Foster of the London School of Economics to undertake a study on the "principles which should underline transport policy in Ireland". Given that key members of the NESC were, or had been, senior civil servants in the Departments of Finance, Economic Planning and Development, this pointed towards high-level misgivings about the national transport situation.

Foster's report was published in 1980. It was critical of cross-subsidisation of the railways from road transport and was sceptical about the future of the railways. It was also critical of the split in government whereby no one department had overall responsibility for transport – with the Department of Transport being responsible for public transport and railway infrastructure while the Department of the Environment was responsible for roads. Foster also recommended, as many had done before, that a Dublin Transport Authority be set up.

A joint committee of the two chambers of the Dáil had also been looking at the financial arrangements between CIÉ and the state and in a report published in 1979 made certain important recommendations, including one which stated that subventions should be paid 'above the line' rather than as offsetting losses 'below the line', as had been the normal practice. This had been a recommendation in the first McKinsey report.

In late 1979 the government decided to appoint McKinsey for a second time to examine the problems of CIÉ. McKinsey reported in 1981 and its main recommendation was to split CIÉ up into three separate companies; a national rail company, a national bus company and a Dublin bus company. Here we were, back where we started – the GSR, IOC, and DUTC all over again! Possibly developments in Northern Ireland in 1967 had an influence on their thinking. The UTA was split into Northern Ireland Railways and Ulsterbus. In 1973 Ulsterbus then took over Belfast's bus services as 'Citybus'.

During the course of the study the decision had been taken to set up a Dublin Transportation Authority (DTA) and this apparently influenced the consultants' thinking regarding setting up a Dublin bus company. As the DTA would have responsibility for the management of Dublin's transport infrastructure and for the planning and development of public transport services for the city, then Dublin Bus could be set up as a separate entity. If that was done, then McKinsey believed that there was a logic in splitting up the rest of CIÉ.

McKinsey could find no further room for major financial improvements in the operation of CIÉ. They believed that the basic problem was that CIÉ had been created around the railway and so had no choice but to support the railway. For this reason they said that the railway had not adapted as radically as required to market needs and so, for the future, it would be better if the railway had to stand alone, separate from road services.

They also stated that since public transport was a concern of the government, then public transport requirements should be decided by government and not by CIÉ. DTA would do this for Dublin Bus and the Department of Transport should do it for the railway and provincial bus services. The consultants believed that if the key responsibility for setting public transport priorities lay with government, then government would be forced into an understanding of the value of public transport and begin to treat it better than they had in the past.

McKinsey also examined whether CIÉ should be retained as a holding company, but decided against it.

The railways were identified as the main problem area for CIÉ but, as in their earlier 1971 report, they saw no great economic advantages in significantly reducing or totally closing down the system.

McKinsey's specific recommendations for the railways were:

- Scrap the freight sundries business since it had become totally financially unviable.
- Close road freight operations, whose main business was sundries.
- Discontinue individual container services, which were only viable when carried on sundries trains.
- Scrap the area structure which would not be needed once the integrated CIÉ was dismantled.
- Centralise all railway operations.
- Improve marketing.

The key recommendations for urban bus operations were:

- Improve operational performance by:
  (a) Introducing OPO.
  (b) Upgrading the fleet with General Motors vehicles.
- Improve service performance by:
  (a) Increasing bus speeds (the main responsibility for this would fall to the new DTA).
  (b) Adapting bus service levels after a full scale route and frequency review.

The key recommendations for rural bus services were:

- Use the standard 54-seat bus only on relatively few high-demand routes.
- Experiment with smaller buses, including privately owned buses, on more lightly trafficked routes.

They also recommended, again, that state subventions should be paid 'above the line'.

CIÉ welcomed the concept of the DTA and

support for 'above the line' payments of subventions. However, they rejected the suggestion that government should take over responsibility for deciding service levels and revenue. They also opposed the break-up of CIÉ, stating that this was an opinion not supported by any facts, and that the integrated approach of CIÉ was regarded elsewhere in the world as an advanced approach to the provision of public transport services.

CIÉ also believed that McKinsey's view of the railways lacked a proper appreciation of both the current value of the network and its long-term potential. They felt that the consultants had not properly quantified either the social contribution made by the railways or its advantages from the viewpoint of energy conservation.

McKinsey's recommendations were not acted upon immediately, but they would later have an echo in the creation of Iarnród Éireann, Bus Átha Cliath and Bus Éireann.

The railway, 1980-83

In 1980 the Minister of Transport informed CIÉ that the government was not prepared to back the proposal that Link Hoffman Busch (LHB) take over some of the Inchicore manufacturing facilities to build coaches for CIÉ and for export. CIÉ therefore submitted a proposal to build its own mainline coaches, using its own staff and using LHB as technical consultants. The government finally gave its approval for the acquisition of new suburban line coaches in 1980, but not for mainline coaches to be built at the Inchicore works.

In 1981 it was announced that CIÉ was to buy 40 two-car EMU sets from LHB for the Howth-Bray electrification scheme. This was about half the fleet requirements for this project and there was some consternation in political circles that they were not to be built at Inchicore. However, later that year the government announced that 124 mainline coaches were to be assembled at Inchicore from kits imported in 'knocked-down' condition. Production began in 1983 and the first of these new Mark 3 coaches went into service the following year.

In the same year, the Minister of Transport announced that a new commuter service to Maynooth would be introduced later in the year along the old MGWR Mullingar line. This was a government and

not a CIÉ initiative, probably for electoral reasons, but because of severe shortages of rolling stock the operation of this service put CIÉ under severe pressure.

By 1983 the first of the new Howth-Bray EMU stock had arrived, new coaches were under construction at Inchicore and new buses were in operation. In addition, for the first time in 15 years, CIÉ's deficit had begun to come down.

Introduction of the DART, 1984

The electrified Howth-Bray suburban service, branded DART for Dublin Area Rapid Transit system, began operating in July 1984. Right from the outset it was popular with the travelling public, with passenger numbers approximately doubling in the first full year of the system's operation from about 6 million to just under 12 million. By the mid-1990s the figure would be 16 million.

However, the feeder bus service that had been meant to complement the DART service did not start on time. It had become entangled in the continuing row between CIÉ and the unions over the implementation of OPO. It was another 18 months before these buses, in their DART livery, entered service. In the first year of services using feeder buses, 2 million additional passengers were carried by the DART.

At the time of the official opening of the DART, the Taoiseach, Dr Garret Fitzgerald, a railway enthusiast who had been a lecturer in transport economics in the past, indicated that he was well aware of the problems faced by public vis-à-vis private transport. He said, *inter alia*:

> Public transport ... is placed in an invidious and economically untenable position by virtue of the fact that one of the most scarce of all resources, road space on the main axes into and out of major cities at peak hours, is provided free of charge to anyone who wants to drive a motor vehicle along it. ... In this way, the relative economics of private and public transport are totally distorted, because private transport is made available at far below the real cost to the community, while the cost of public transport is enormously increased by the congestion of the roads by under-utilised private vehicles.

The 'Building on Reality' report, 1984

In 1984 a new chairman, CT Paul Conlon, took over from Liam St J Devlin. He was an accountant with a previous record in turning around businesses which were in serious financial difficulties. It was hoped he could do the same for CIÉ.

He began to look critically at CIÉ's organisational structures. He informed the government that he was attracted to the McKinsey recommendation of setting up three operating companies for rail, provincial bus services and Dublin bus services. However, a level of co-ordination and integration would be maintained between these three companies through transforming CIÉ into a holding company. In this form, CIÉ would be small, tightly organised and provide a limited number of support services to the three operating companies.

He initially recommended hiving off road freight into a fourth company. This also had echoes of the road freight division of the UTA being set up as Northern Ireland Carriers in 1965. However, the government preferred road freight to remain as part of the railways, so the proposal for the road freight company was dropped.

The restructuring of CIÉ was announced in the government's response to McKinsey's *Building on Reality*, published in late 1984. As well as proposing the setting up of three operating companies, it was stated that there would be no new substantial investment in the railways, that rail sundries and road freight services would be terminated from January 1986, unless they had turned a profit in 1984-85, and that there would be an increasing emphasis on the development of bus services. A proposal to establish the long-proposed Dublin Transportation Authority was also announced.

In 1984 'Inter-City' branded trains, using new Mark 3 carriages, upgraded track and the new CTC signalling system, were introduced. The CTC was able to deliver a combination of increased safety and speed on the main lines.

The Annual Report for the year commented that the payment of the subvention 'above the line' was:

> ... a recognition by the Government that a National Public Transport system, combining road and rail services, is essential in the public interest and that certain infrastructure and social service obligation costs should be properly met by the state.

In 1985 and 1986 improvements in passenger numbers took place. Rail passenger numbers grew by

DART EMU set at the 'Inchicore 150' celebrations.   Author

40%. Those on non-electric suburban services grew by 38%, mainly as the result of cascading better quality coaches from mainline to suburban services, as a result of the introduction of the new Mark 3 stock. Traffic on Dublin bus services grew by 6%, those on provincial services by 3% and Expressway by 8%.

Returning now to the re-structuring of CIÉ, the three operating companies were to be called *Iarnród Éireann* (Irish Rail), *Bus Éireann* (Irish Bus) and *Bus Átha Cliath* (Dublin Bus). They would be set up to give each company as much autonomy as possible. Each company would have its own managing director and board. The holding company, CIÉ, would retain a limited number of functions to ensure overall cost-effectiveness and operational efficiency. These would include, from within the sums provided by the government, the financial targets and allocations for the three operating companies, and CIÉ would guard against 'reckless or ruinous' competition between the operating companies. Competition between the subsidiaries would be monitored by CIÉ, but not discouraged.

The main area of contention when planning the new structures was whether Iarnród Éireann's mainline services could survive if Bus Éireann was to fulfil its own mandate by competing aggressively. It was agreed by the government, as a guard against this happening, that the majority of directors would be common to both Bus Éireann and Iarnród Éireann. It was hoped that this would ensure that the boards of both companies would take full account of the interests of the individual companies.

De-centralisation,1987

The subsidiary companies were incorporated in January 1987 and began trading the following month. Shortly after the establishment of the three new operating companies, the chairman, in a statement to staff, made three key points:

- The purpose of the re-organisation was to give the whole CIÉ organisation a much stronger customer and market focus.
- The total market for transportation was growing and not declining.
- Their main competition came from the private car and not private bus operators.

In 1985 the government released a consultative (Green) paper, on future transport policy. In this it was stated, with regard to the railways, that because of the investment decisions already taken by government, it had been decided to retain the railways in the medium term. This would allow time to consider whether:

- The existing railway network should be retained in the long term.
- Retention should be based on a reduced network.
- The network should be completely closed.

Retention of a railway network would be dependent on continuing state financial support. It was hoped that the trade-off between the costs and benefits of these alternative strategies would, as a result of the consultative paper, be debated further. Conclusions drawn by the government from this debate would be incorporated into a White Paper to be published in 1987.

It was clear from these statements in the Green Paper that the government was giving no long term commitments to the railways; in fact the opposite seemed to be the case.

In 1986 two further changes took place. Firstly, large scale introduction of OPO buses commenced in Dublin and, secondly, the canals were removed from CIÉ to the Commissioners of Public Works. The removal of the burden of running the canals was something that CIÉ had long requested.

Iarnród Éireann

In early 1987, CIÉ made a case to government for the construction of 18 railcars plus the modification of some Mark 3 coaches so that they could be operated with the railcars. This would have allowed CIÉ to reduce by an equal number the new locomotives needed to replace the ageing 001-class Metro-Vick locomotives. If agreed, it would also have dented the government's 'no new investment in the railway' policy, referred to above. The project was not approved and CIÉ had to wait seven more years for its new 'Arrow' railcars and 201-class locomotives.

Iarnród Éireann (IÉ) continued, as best as possible within the prevailing financial constraints, to extend the CTC signalling system, to upgrade track and complete the Mark 3 coach build programme.

The last 24 Mark 3s had been adapted for suburban work by being modified to suit push-pull work, using 121-class locomotives as power units. They would be used on outer suburban work on the Dublin-Drogheda and Dublin-Maynooth lines.

In order to reduce the number of timber-framed coaches in service, three 80-class railcar sets were leased from NIR for three years while, in 1989, 15 British Rail Mark 2a coaches were bought from Vic Berry, a Leicester railway scrap dealer, and were extensively refurbished in Inchicore. They came into service in 1990.

Figures produced in 1990 showed that 'Inter-City' services were carrying more passengers than at any time since the foundation of CIÉ and looked set to grow further. On seven inter-urban corridors, surveys showed that although travel by car was top, with 68% of the journeys; rail was second with 24% and buses third with only 6%. Air travel claimed 1%. On the Dublin-Cork corridor, rail carried 33%. This corridor accounted for a quarter of the total number of passengers using rail travel.

In 1990 IÉ could point to having kept fares steady since 1986, to a reduction in state grants due to better market performance and improvements in productivity and with a network that still served most major centres of population. It had 24% of inter-urban passenger movements and 17% of long-distance freight. The contribution of the DART in helping reduce Dublin's traffic congestion was clear and the system was generally regarded as a major success.

In 1987 it was finally announced by the government that the electrified system would not be extended westward to Maynooth or northwards to Ballymun. The proposal for a city-centre underground link was also dropped. CIÉ was asked for new proposals for public transport developments to the west, based on buses and diesel services on existing lines. CIÉ's recommendations were:

- The opening of four new stations on the Connolly-Maynooth line.
- The introduction of commuter services between Clondalkin and Pearse with the re-opening of eight stations.
- An expansion of the 'Localink' bus network which had been introduced in the Tallaght area in 1987.

- An expansion of the bus feeder services to complement the rail extensions.
- The development of a busway along the old Harcourt Street line.

CIÉ again asked for new railcars for their commuter services, actual and proposed.

The four stations on the Maynooth line were opened in 1990. The commuter service on the Clondalkin line would be initiated in 1994. The Harcourt Street proposal become part of the 'Luas' tramway proposal in 1994.

In CIÉ's 1990 Annual Report, the company protested at the significant road bias in the European Community Regional Development Programme for Ireland, where only £36 million had been asked for rail-related projects, whereas £623 million was earmarked for road developments. However, the following year the company's Annual Report could point to some redress in this imbalance. It stated:

> Since then there has been increased public awareness of the deteriorating condition of much of our rail infrastructure and the need for major additional investment is now widely recognised. Throughout Europe, governments and national authorities recognise the essential economic, environmental and social value of railways and are now actively pursuing invest-ment in them with the support of substantial EC funding.

The reason for this change of tone was due to the EU's decision to contribute to the development of the Dublin-Belfast line as a major European rail link. This was supported by both the Irish government and by the Northern Ireland Office of the UK government. The 1991 Annual Report went on to say:

> Government support for this project is regarded as a flagship commitment for railway develop-ment. It is confidently expected that similar support will be forthcoming for other radial routes.

In fact, there does seem to have been a sea change taking place about this time in the government's attitude to the railways. There were probably several key factors responsible for this. Firstly, IÉ was producing very favourable financial results. Secondly, the government was being influenced by the obviously strong support for railways by other governments on mainland Europe. Thirdly, they began to see how investment in the railways could help contain the ever-escalating demands for investment in roads to ease traffic congestion. Finally, railways were more environmentally friendly and more energy efficient and these were factors which were becoming evermore politically important.

The idea that the maintenance of the railway infrastructure be separated from railway operating costs, recommended in the Milne Report in 1948, had now received EU support. In 1991, an EU Directive required that henceforth, in railway accounts, maintenance of infrastructure should be presented separately from operating costs. It also provided that any authorised operator should have access to any member state's rail infrastructure, on the grounds that the permanent way was to be regarded as a public asset, rather than the property of the operator. In Ireland, this last provision would not be of any significant practical importance, but the first one would allow CIÉ to focus attention on the costs of maintaining the permanent way.

In 1992 the government approved the purchase of 17 railcars for commuter services. It was also decided to allow the replacement of the 001-class locomotives and a contract was signed with General Motors for ten new 201-class locomotives. This order was soon increased by a further 22 locomotives and further sums were approved for an additional ten diesel railcars and other related investment in commuter services. This was the single biggest investment in railway rolling stock since 1953, when CIÉ had dieselised its railway operations.

The National Development Plan, 1994-99

In October 1993, the 1994-1999 National Development Plan was launched. This plan was to be assisted by £8 billion of EU Structural, Regional and Cohesion funds. With regard to the railways, during the period £185 million would be invested in upgrading the main lines from Dublin to Galway, Waterford, Cork, Limerick, Tralee, Sligo and Belfast. This was to include track and signalling renewal, plus new rolling stock. In addition, IÉ was to spend an additional £90 million on improving the Dublin-Westport/Ballina and Dublin-Rosslare lines.

The plan pointed out that Dublin would also benefit from ten Quality Bus Corridors (see below) and the upgrading of suburban rail services. A figure of £200 million was also allocated for a Light Rapid Transit (LRT) system for Dublin.

The year 1994 saw the inauguration of the new railcar service from Heuston to Kildare, rather than from Pearse to Clondalkin, as originally proposed. The ten new railcars used on this service were built by the Tokyu Car Corporation of Yokohama, Japan. They were designed in the same general style as the DART railcars and were marketed under the brand name 'Arrow'.

The same year also witnessed the delivery, by air, of IÉ's first 201-class locomotive from GM's plant in Canada. The balance of the initial order for ten locomotives was delivered before the end of the year. Another 20 arrived in April 1995.

On the handing over of the first of the new 201-class locomotives, on 24 June 1994, Iarnród Éireann introduced their new Irish-designed logo based on the initials of the Irish version of the corporate name, 'IÉ' as opposed to the previous logo 'IR', which was based on the English language version of the corporate name, 'Irish Rail'.

Two further 201-class locomotives were delivered to NIR to be used on the new joint IÉ/NIR Belfast-Dublin 'Enterprise' service. Trains and locomotives on this service would use a common brand name and livery. They would be made up of new dedicated coaches, to be built by De Deitrich of France.

As the renewed search for a solution to Northern Ireland's political problems developed, traffic on the Belfast-Dublin corridor increased, so that a shortage of capacity on the new 'Enterprise' service rapidly became a problem.

Developments in 1995-1999

In May 1995 the Minister for Transport announced that the DART would be extended after all, but south to Greystones. Work would be carried out with the assistance of European Union funding. The following month it was announced that the DART was also to be extended north to Malahide and two new stations would be built, at Fairview and at Barrow Street, between Pearse and Landsdowne Road. It was planned that the Greystones extension would be opened in 1998 and that to Malahide by 1999. New 'Park and Ride' facilities would also be put in place.

In August 1995 the final report of the Dublin Transportation Initiative (DTI) was released by the government. This £1.2 billion plan was aimed at solving Dublin's traffic problems. The DTI outlined a transport strategy for the Greater Dublin area up to the year 2011, including the provision of a LRT system, Quality Bus Corridors, extension of the DART, as mentioned above, and stricter enforcement of traffic laws. It was also announced that the DART extensions should have priority and it recommended that, as well as the two new stations, existing stations should be upgraded and five additional two-car EMU units would be bought. It was also proposed to link Tara Street station with a new public transport interchange.

A contract for the design and installation of these extensions to the DART network was let in 1998 to Adtranz of Germany. Work on upgrading stations on the DART system began in mid 1997 and a new station at Clontarf Road was officially opened in October 1998. The new station was opened near to the site of the original Clontarf station, which closed in 1956. A limited service between Bray and Greystones was planned for February 2000 with a full service by June 2000.

With regard to non-electrified suburban services, the priorities of the DTI would be the upgrading of the line between Clonsilla and Maynooth, doubling the track, modernising the signalling and purchasing new rolling stock. Plans were announced for two new stations, plus the re-opening of Drumcondra station on the former GSWR line. Drumcondra was in fact re-opened in March 1998.

Outer suburban services, particularly Dundalk-Arklow, would be upgraded, as required by increased passenger demand. As part of the latter, work on upgrading Drogheda station, to handle both the improved cross-border services and increased suburban traffic, was begun in 1997 and completed in 1999.

In September 1996 IÉ announced plans to rebuild the ferry terminal at Dun Laoghaire station in such a way as to encourage ferry passengers to go directly to DART or other rail services. The modernised station was officially opened in May 1998.

In 1998 a proposal was announced by the Fine Gael party and Dublin Chamber of Commerce for the location of a new Dublin 'central' station in the North Wall area. This proposal included the construction of a new link line between the station and the Rosslare line, making a connection with the latter at Barrow Street, and the upgrading to passenger standards of the existing lines from the North Wall to Newcomen, North Stand and East Wall junctions. If built, this new station would probably result in the closure of Pearse, Tara Street and Connolly, the three busiest suburban stations on the system. The proposed new station is much further from the city centre than the existing stations and, even if connected to the city centre by a bus or tram link, would be far less convenient for most commuters than the present stations.

In 1998 and 1999 major rebuilding of both Heuston and Connolly stations was undertaken. In the case of Connolly, this amounted to a virtual rebuild, though preserving the historic frontage of the old station. The new structure includes an office block over the station.

Another improvement, which was introduced in this period, was the provision of a special bus service between Connolly and Heuston stations using single-decker vehicles carrying a variant of the DART livery.

In 1996 IÉ invited potential tenders for a fleet of new diesel railcars. These were similar to the 'Arrow' sets but were more spacious. An order was placed in 1996, with GEC-Alsthom Transporte in Spain, for 27 units to be built near Barcelona. Inevitably these new 2700-class railcars were dubbed 'Sparrows' by railwaymen. Deliveries commenced in April 1998 and were completed during 1999.

Another order, for ten new DART vehicles, consisting of five driving motor and five driving trailer vehicles, was placed with GEC-Alsthom in 1998. These vehicles were also built at their Barcelona factory and delivery commenced in late1999.

The 1998 Irish Rail Annual Report, published in July 1999, showed a continued growth in revenue, although the increase was partly offset by a rise in expenditure. There was a surplus after grants of almost £4.5 million for the year. Passenger journeys rose to an all-time high of 32.1 million, up 8.2% on 1997. On Inter-City routes, journeys rose for the third successive year, reaching 9.83 million. Passenger journeys on the DART system was 19.71 million and on non-electrified suburban services, 2.56 million. The number of passengers travelling on the DART is now double that when the service began.

Bus Éireann

Bus Éireann started its independent existence in a

New Spanish-built 2700-class 'Arrow' railcars at Inchicore, shortly after delivery in 1999.   Author

highly competitive market and its early financial results were poor, making a loss of £4.2 million in 1987. Expressway services were profitable and cross-subsidised provincial stage carriage and Supabus services which were not. Bus Éireann's passenger numbers grew by about 6% per annum each year between 1987 and 1993. Its tour business, mainly day tours and private hire work, also grew rapidly and moved from loss to profit by 1993.

CIÉ Tours International, a separate subsidiary of CIÉ, also began to show a rapid improvement. Much of this was the result of a combination of improved marketing and better control of expenditure.

In 1993 Bus Éireann earned a surplus of £3.4 million after a state grant of £4 million towards meeting the costs of the social role of the company's services.

The new company was free to source its own buses, mainly because of requirements now imposed on Ireland when it had joined the EEC. By 1993 the 202 strong 'Expressway' fleet was a mixture of Leyland, DAF and Volvo chassis with mainly Cummins engines, also to be standard in the 'Arrow' railcars, and bodies by Van Hool, Plaxton, Duple and Alexander.

In 1994, 20 coaches, Volvos with Caetano bodies,

were bought. The provincial city fleet was augmented in 1993 by ten high-capacity DAF single-deckers with bodies by Alexander of Belfast. These were used to replace ten double-deckers.

Bus Éireann has ordered 22 Mercedes Ciatarro single-deckers to further improve services in Cork and Limerick. Pending delivery of these, a number of Plaxton and East Lancs bodied Dennis DART SLFs have been hired, to speed up the phasing out of older step-entrance buses, such as the KC-class GACs. Some of the KC-class buses are being converted to single door school bus format, becoming the KCS-class.

The last double-decker in ordinary service with Bus Éireann, a KD-class Bombardier/GM vehicle, was withdrawn late in 1999, thus finally bringing to fruition the plan announced by CIÉ in 1976, to replace all provincial city double-deckers with high capacity single-deckers.

Bus Éireann is due to receive a further large batch of Volvo B10 full sized low floor single-deckers in the year 2000, to augment 11 received in 1996-97, thus speeding the demise of older buses such as the 1983 GAC city buses.

The school bus situation was a problem, as the government was not prepared to fund the acquisition of new vehicles. As a result, Bus Éireann collected a variety of second hand vehicles, 101 from the UK Ministry of Defence, 104 from Ulsterbus/Citybus, 130 New Zealand-built ex-Singapore Volvos and 95 converted ex-Expressway vehicles.

Bus Átha Cliath

Bus Átha Cliath was serving a city and county whose population had increased by 45% over the previous 30 years, but in spite of this, between 1957 and 1987, passenger numbers had fallen from 249 million to 167 million. Dublin city services had once cross-subsidised the railway, but by the 1970s they had moved into loss. The two main reasons for this decline: firstly, the growth in ownership and use of the private car, allied to the slowness of government and civic authorities in tackling the consequent traffic congestion problems; and secondly, the terrible industrial relations record of Dublin bus services.

Although Dublin Bus periodically checked its route network and made route changes from time to time, it rarely came up with any new service concepts. It had drawn up plans for a rapid transit system for Dublin, including bus lanes and electrified commuter railway lines and a city centre underground and transportation centre. However, these still lay with government, which had failed to decide on them one way or the other. All buses bought since the 1960s had been capable of one-person-operation but, until 1986, unions had successfully blocked progress on this front.

However, by 1988 75% of Dublin's bus services had been converted to one-person-operation and by 1990 staff numbers had been reduced by 25% as a result of the introduction of OPO. As part of the OPO initiative, pre-paid tickets were sold through retailers and ticket cancelling machines were fitted to buses.

Minibuses were introduced to serve the Tallaght, Blanchardstown and Finglas areas. New improved specification double-deckers – Leyland Olympians, with bodies by Alexander of Belfast – were ordered and entered service in 1991.

Although the financial results of the early years of the company were very poor, passenger numbers had started to increase again. By 1991 revenues were up by 12% over the previous year and costs had fallen.

Also in 1991 Bus Átha Cliath began an experiment to determine whether double-deckers were in fact the best way of providing urban services. A low frequency service of double-deckers was replaced by a high frequency mini-bus operation. Drivers recruited for the mini-bus operation were paid 20% less than regular drivers. The service was marketed using the 'City Imp' brand name. On the experimental route on which the concept was tested, passengers numbers trebled immediately. Over the next couple of years, 'City Imp' services were introduced in other parts of the greater Dublin area. Another 1991 innovation was the 'Nite-Link' service providing late night transport between city centre and suburban centres.

Another major experiment was the Quality Bus Corridor (QBC). This was based on the idea of trying to identify and adopt for buses, some of the advantages of a light rail system. The main features of the QBC system were:

- **Route alignment:** This was the fixing of a definite path for each bus route, one which would be popular and well known.
- **Bus frequency:** Experiments, like the City Imp one described above, showed that a seven or eight minute frequency best appealed to the public. If people could be sure of a bus at frequent intervals they would be more likely to use them.
- **Bus design:** Buses should be clean, easy to board, quiet, and attractive.
- **Staff selection and training:** Staff should be able to relate well to the public, to be able to handle cash and be prepared to work irregular hours.
- **Infrastructure provision:** Bus stops should be provided with attractive, well kept shelters, with a seat and lighting, and in some cases even a telephone.
- **Traffic Priority:** Bus priority should be granted by the civic authorities allowing easier and faster bus flows through traffic congestion.

The first QBC, route 39 to Blanchardstown, was inaugurated using the new grey and blue liveried 'Cityswift' single-deckers in June 1993. The following year more QBCs followed to Ballyfermot, Finglas and Bray.

By the end of 1994, Dublin bus services had undergone a major image change with new improved buses, new liveries and new products. The 1993 financial results were the best for 20 years, although the company was still showing a deficit.

In the CIÉ Annual Report for 1998, £5.5 million was allocated to Dublin Bus for the purchase of additional vehicles, and £16 million to the Dublin Transportation Office for traffic management grants.

Dublin Bus took delivery of 195 Olympian/ Alexander double-deck buses during 1999. Originally only 70 were ordered, but extra funding became available from the European Union and Dublin government, due to delay in starting work on the Luas light rail system.

During 1999 Dublin Bus had six low-floor double-deckers on a six month trial – two DAF DH250s, two Volvo B7Ls and two Dennis Trident II/ Alexander ALX400s. Dublin Bus has ordered 100 Volvo B7Ls with Alexander ALX400 dual-door bodywork, for delivery in 2000. It is hoped to order a further 85 double-deckers during the year.

Following on from the delivery of 20 Wright-bodied Volvo B6BLs in 1999 (see page 124), a further 20 are on order. Another trial of articulated single-deckers will take place in 2000, with a possible order for 20 such vehicles, for use on short but well-patronised routes such as the Heuston to Connolly shuttle.

Light Rapid Transit (Luas)

As outlined earlier, studies carried out in the early 1970s had recommended the construction of a Light Rapid Transit system for Dublin, with electrified lines from Howth to Tallaght and Bray to Blanchardstown, including underground sections in the city centre. The Howth-Bray electrification, approved by CIÉ in 1977, was in fact to be the first stage of such a system. This project was given government approval in 1979. However, the government did not support the rest of the scheme at this time. By the early 1990s, with Dublin traffic conditions becoming increasingly difficult, government attitudes had changed.

In October 1993 the National Development Plan 1994-1999 was unveiled in Dublin. This plan, which was to be assisted by EU funding, included a figure of £200 million for a Light Rapid Transit system for Dublin. At this stage there were no details given of the proposals and the minister said that these could not be finalised until the Dublin Transport Initiative had been completed and given to government.

In mid-1995 the government indicated that, with

regard to the LRT project, it expected to see the first two lines, to Tallaght and Dundrum, in place by 1999. The third link, to Ballymun and the Dundrum to Cabinteely link, would follow later.

The core network, costing £200 million, would include the routes from the city centre to Tallaght, Cabinteely using the old Harcourt Street line, and Ballymun. At a later stage, the DTI recommended a fourth route to Finglas via the disused Broadstone-Liffey Jct. Line, and a fifth linking Dublin Airport as an extension of the Ballymun line. It was estimated that the total cost of the project would be about £410 million.

In December 1995 the government provided details of the proposed LRT system and an implementation timetable. According to this both lines would come into operation in 2000. 'Luas' (the Irish word for speed) was adopted as a working name for the system.

It was stated that the journey times would be 30 minutes to Tallaght and 20 minutes to Dundrum. The first phase would include 21 km of double-track and 32 stations. Up to 29 trams would be used during peak hours, with a service frequency of six minutes. It was announced that the European standard railway gauge of 4'8½" would be used, rather than the Irish gauge of 5'3".

Tenders were sought in September 1995 and three were short listed – Siemens, Bombardier Eurorail, and GEC Alsthom. The preferred supplier is GEC-Alsthom.

In May 1997 CIÉ submitted its application for a Light Rail Order (LRO). This was accompanied by the Environmental Impact Study. It was expected that the Public Enquiry would be established by mid-1997 and the LRO granted early in 1998.

CIÉ was expected to begin the public consultation process for the Ballymun line and the extension of the Dundrum line to Sandyford by late 1997. The government approved a mid-1997 start on planning and design work for the Sandyford extension and allocated state funds, plus the matching EU finance for the cost of designing the Ballymun extension.

However, in mid-1997, the Minister for Transport appointed consultants WS Atkins to re-examine the plans for the Luas system, in particular the question of putting the lines underground in the city centre. Although the consultants, when they reported back in

April 1998, recommended that they supported the CIÉ preference for on-street-routing, the government decided otherwise and radically altered the Luas scheme to include an underground section in the city centre.

Under the new proposal there will be two lines, one running east-west, linking Tallaght to Connolly station and the other running north-south, connecting the airport to Sandyford. The east-west route will follow the CIÉ proposed alignment from Tallaght, via Heuston station to O'Connell Street. Then, instead of turning south to Sandyford, it will continue east to Connolly station. The north-south alignment will follow the CIÉ route from Sandyford to St Stephen's Green, then go underground for 1½ miles and emerge again at Broadstone station. From there it will run to Ballymun and then the airport. The journey between Sandyford and St Stephen's Green will be about 20 minutes. A new tram depot will be built at Sandyford Industrial Estate.

Future proposed extensions include lines from Connolly station to the docklands re-development area, from the airport to Swords, from Sandyford to Cabinteely and from the Naas Road to Clondalkin. The Dublin Transport Office has drawn up a plan to later extend the system to serve Stepaside and Carrickmines.

However, these routes could yet be revised in the light of further studies, public appeals, political considerations and geological factors.

As a result, the cost of the project will be considerably increased, from IR£240m to IR£400m. The start date has now slipped to the autumn 2000 and will take about three years. Existing European Union funding for the project has been lost and a fresh application will now have to be made. However, the Irish government seems determined to push forward with the project and detailed planning work is continuing. The size of the initial fleet of vehicles has been increased from 29 to 32 vehicles.

Developments at the end of the century, 2000

The development of public transport services, both road and rail, in the Republic of Ireland look very positive as this book goes to press.

Railways

The Minister for Public Enterprise, when announcing a £461 million rail safety programme, also announced a government commitment to the retention of all the existing railway network. She stated that this was to end the uncertainty created by decisions announced in the 'Building on Reality' report of the mid 1980s.

The Department of Public Enterprise announced the commencement of major planning processes, for the medium to long term, covering the Dublin and Cork regions and is based on making a more intensive use of existing rail corridors. Possible measures being looked at include:

- Provision of additional DART and Arrow rolling stock.
- Increasing DART and Arrow trains to eight car formations.
- Upgrading the Heuston-Connolly line, including the provision of additional stations.
- Upgrading Mallow-Cork-Cobh services.
- In the Cork area, the possible re-opening of the Youghal line as far as Midleton.

In addition, consultants advising the government on the way forward on transport matters over the period 2000-2011, have stated that future developments must be based around public transport. They have put forward a number of rail-based proposals for the Dublin area for consideration, including:

- Provision of a line across the Liffey to the east of the existing link line, to increase rail capacity in Dublin city centre.
- A heavy rail line to the airport, possibly from the Maynooth line. This could also serve the Swords area.
- The diversion of main line services on the northern line, via the airport, possibly to a new terminus at Broadstone. This would also increase capacity on the existing line for additional suburban services.
- Separating suburban and inter-city services by quadrupling track.
- Reinstating the inland route from Clonsilla Junction to Navan. The routing of this line through Blanchardstown town centre, together with the possibility of a link through Dunshaughlin and Ashbourne to Swords, will also be examined.

New Bus Eireann Volvo, with body by Wrights of Ballymena, on display at the ITT bus rally in 1999.   Author

The government has approved the subjection of these short and long term proposals to detailed costing and feasibility studies.

Buses

The managements of both Bus Éireann and Bus Átha Cliath are pressing for funding to reduce the average age of their bus fleets to five years and, in Dublin, to complete the network of Quality Bus Corridors. As noted earlier, the introduction of the new 'City Swift' branded buses, using QBCs, has led to significant increases in patronage on the routes where they have been introduced, whereas patronage has continued to fall on routes operated in the traditional manner.

Both companies are aiming to provide services which are as time-certain as trains. In fact the aim is to give bus customers a DART-type bus service on the roads of Dublin. This can only be done if there is dedicated road space for the buses, and so the QBCs must be completed.

The target in provincial cities is for more, newer buses, operating on re-vamped routes and timetables, providing a high frequency, reliable service. Bus Éireann's management are also demanding dedicated road space for their buses in provincial cities or, at the minimum, priority rights-of-way and are working closely with local government to try and provide this.

In addition to speeding up travel times and improving the quality of the vehicles, increasing emphasis is also being placed on improving information to passengers and providing first-class facilities both before and after journeys. It is also planned to up-date ticketing with the introduction of a 'smart card' ticketing system.

Luas

The plans for the Luas system have been revised several times and the relative merits of street or underground running in Dublin city centre have been examined in depth. The current (March 2000) proposed network is for one line going underground, while the others stay on the surface. Details for much of the system are still sketchy, but proposals for the lines from Tallaght, Sandyford and Connolly stations (Lines A, B and C) to the city centre are now available. Line D (the 'Northside' line) and its underground link through the city centre will be the last of the initial phase to be determined.

A favourable report from the public inquiry into Line A (Tallaght) was published in December 1998. A public inquiry into Line B (Sandyford) reported in July 1999 and recommended construction of this line. The public inquiry into the short Line C to Connolly station was published in January 2000. It rejected the proposed terminus at Harbourmaster Place, which will now have to be relocated. Initial on-site work has commenced on Line A, with construction due to start in spring 2000 . Trams will first operate in late 2002. The four Lines in Phase 1 should be completed in 2005.

Models of the Luas vehicles were unveiled at the end of February 2000 (see page 125 lower) and it is planned that the first will arrive in 2001. Twenty units have been ordered for the Tallaght line, at a cost of over £IR20 million. A further 14 will be ordered for the Sandyford line, with an option of another six for the Abbey Street – Connolly station route. Each of the silver coloured units will seat 60 people and carry a total of up to 240 passengers.

CIÉ have issued tender documents to firms interested in bidding for the design, installation, testing and commissioning of all trackwork, overhead line equipment, power supply equipment and control systems for the three lines. The value of the contract is about £IR100 million.

Line A: Abbey Street toTallaght .

This line will run for 14 km from the city centre, through the north inner-city, crossing to the south of the river at Heuston railway station, before serving St James, Rialto, Drimnagh, Blue Bell, Red Cow, Cookstown and Tallaght. There will be a depot at Red Cow. This will be the first line built, with completion scheduled towards the end of 2002. Of this route, seven km is 'on street', the remainder being on dedicated alignments or on the central reservation of the main Naas Road (N7). The original terminus was to have been in Middle Abbey Street but, with the Line C plan now finalised, this has been replaced by a stop in Lower Abbey Street, where the two lines join. Twenty low-floor trams will carry 2,800 people per hour in each direction with five-minute headways at peak times. The journey time will be about 38 minutes.

Preliminary work on this line began on 19 August 1999, six weeks ahead of schedule, with the raising of the ESB electricity wires in Cookstown industrial estate to allow the construction of the overhead line equipment for the tramway.

Line B: St Stephen's Green to Sandyford.

This eight–nine km line will run from Sandyford to the city centre, serving Balally, Dundrum, Milltown and Ranelagh. It will largely follow the route of the closed Harcourt Street railway line, but there will be a one kilometre section of 'on street' running along Harcourt Street and St Stephen's Green West in the city centre. A depot will be located at the Sandyford terminus. Thirteen low floor trams will carry 3,000 people per hour in each direction with five-minute headways at peak times (15 minutes off-peak). The journey time will be 22 minutes. Eventually trams on this line will go underground through the city centre, but the precise location of the tunnel has yet to be decided. The line will open in June 2003 though the section on reserved track between the Grand Canal and Sandyford could be opened as early as 2002.

Line B extension: Sandyford to Carrickmines and Cabinteely.

The plan for the extension of this line to the south, would take it away from the old Harcourt Street alignment at Sandyford industrial estate, through Stepaside and Ballyogan to Carrickmines, where it would rejoin the old railway to continue to Cabinteely. It is possible that this extension could be built before line D, due to the ambitious development plans for the area it would serve. These plans might include a levy to part finance the line.

Line C: Abbey Street to Connolly station.

This short route is designed to link Line A (and the city centre) at Abbey Street Middle with Connolly mainline train station and Busaras, the provincial bus station. The route proposed by CIÉ was to run along Abbey Street Lower, Beresford Place and Store Street (Busaras) to a terminus at Harbourmaster Place, behind Connolly station. The public inquiry rejected the proposed terminus at Harbourmaster Place, which will now have to be relocated. Although behind schedule, compared to Line A, this short line could become operational more or less at the same time.

Line C extension: Connolly station to the Docklands.

A future extension into the Docklands development, including the proposed National Conference Centre, is planned. This extension could possibly terminate at the Point Theatre. There is concern that the proposed Luas alignment will have to be single track for some of its length, compromising service frequencies. A new bridge over the Royal Canal would be required and a further problem is that it must be high enough for the canal to be navigable.

Proposals for the National Conference Centre include the provision of a tram stop at Mayor Street. This would also provide an interchange with a station on a proposed new DART suburban railway line. This new DART line would connect with the existing line at Barrow Street, across the river. Currently there are discussions as to whether it will be underground or overground.

Line D: Broadstone to Ballymun and airport

This, the so-called 'Northside line', would form a northern extension to the underground section. Public consultation on possible routes is ongoing. Two main options have been identified, with several variants. However, the most likely route would be to follow the old Broadstone railway cutting as far as Liffey junction, then on to Glasnevin, Ballymun, the M50 interchange and Dublin Airport. Options include longer underground sections to Glasnevin or Drumcondra and a route via Dublin City University.

Underground section: St Stephen's Green to Broadstone.

Exploratory drilling was completed at the end of 1999 to test the feasibility of this underground link, connecting lines B and D through the city centre. The consultants have concluded that the underground tunnel is possible and suggested various possible routes to the government.

Northern Ireland 1945-2000

The UTA, 1948

After World War II, mindful of the failure of the pre-war arrangements involving the setting up of the NIRTB, the Northern Ireland government started to grapple with the whole transport issue in the province again.

In 1946 it issued a White Paper on public transport in Northern Ireland. As a result of this, and in order to avert a catastrophe for the area served by the BCDR, which had made known its intention to cease trading, the necessary legislation was passed quickly. Hence the creation, in 1948, of a new state-owned body to be called the Ulster Transport Authority. This would, initially, take over the assets of the NIRTB and the BCDR, on the understanding that the latter's network would be drastically pruned and appropriate road substitution put in place. Take over of the LMS (NCC) would follow in 1949.

Because of their cross-border nature the UTA would not take over the assets of the GNR, SLNCR, CDR, or the LLSR. Its job would be to unite the railways under its control, by now in a desperate financial state, with the Road Transport Board, into a single, integrated transport unit. Of the small operators which had survived the establishment of the NIRTB, only the Erne Bus Service still survived, and would continue to do so until 1957.

The UTA began to combine the administration of its road and rail operations as a fundamental aspect of its approach to co-ordinating road and rail services. Maintenance and construction were re-organised to avoid duplication. Some road depots were closed and their facilities transferred to railway stations. A common timetable was published showing the road/rail connections and a through ticketing system was established. Fares were standardised over the road and rail systems. The disadvantage here was that, as the level of fares was dictated by the higher operating costs of the railways, the bus fares were probably higher than they would otherwise have been.

Common road/rail design and engineering facilities were established at the Authority's main engineering complex at Duncrue Street in Belfast. Interchangeability of parts between buses and the Authority's new multi-engined and multi-purpose

The former BCDR station at Bangor, with a line up of ex-NIRTB buses in early UTA days. This station was demolished in 1999 and replaced in 2000 by a new Translink combined bus and train station.

Author's collection

A 1953 line up of new UTA MED trains at Queen's Quay station, Belfast.

Author's collection

diesel railcars was achieved through co-ordinated design and procurement policies. The programme of development and use of multiple-engined diesel railcar trains was in many ways advanced for the time, the new trains having, amongst other novel features, automatic doors operated by the guard.

At its formation, the UTA operated over 2,500 miles of road services, owned 282½ miles of standard gauge and 42½ miles of narrow gauge railway, 897 single-deck and 111 double-deck buses, 93 steam locomotives, five diesel locomotives, four railcars and 1,002 lorries. It also took over the operation of the railway owned hotels and other catering facilities throughout the province.

The cull of the railways, 1950-59

However, the UTA's approach to road-rail co-ordination was not appreciated in many parts of the province, when it began to close large parts of its newly acquired rail network. Although the closures might have been expected in the higher reaches of transport and government circles, they were not expected by the general public, who were not pleased by the UTA's new practical definition of the word 'co-ordination'.

Early in 1949, less than six months after the its birth, the Authority gave notice of its intention to close the entire ex-BCDR system, except for the heavily used Bangor commuter line. These closures were swiftly implemented in early 1950.

In 1950 the UTA published its intention to close large chunks of the ex-NCC system. The lines affected were the Ballymoney to Ballycastle narrow gauge line, the Larne Harbour to Ballyclare freight only narrow gauge line, the Kingsbog Junction to Ballyclare freight only line, the Cookstown Junction to Cookstown line, the Magherafelt to Draperstown line, the Magherafelt to Macfin line and Limavady Junction to Dungiven line. Services on these lines were withdrawn later in the year, except for freight services which were retained until 1955-56 on the Limavady Junction to Limavady section, the Cookstown line, and the Derry Central line, (except for the stretch from Cookstown Junction to Kilrea, which remained open for freight until 1959).

In spite of the recommendations of the pre-war McClintock report, no rationale had been established for capitalising the new Authority, which was given

wide borrowing powers. By the end of its first three years, it had accumulated a deficit on its revenue account of £894,406.

As a result, the Minister of Commerce asked the Transport Tribunal to hold an enquiry, which reported back in 1952. It found that the UTA's financial problems resulted from increases in wages, taxes and general overheads. It recommended that the government liquidate the accumulated deficit, assume liability for further deficits within a stated period, during which the Authority would become self-supporting. It also recommended that the main line railways should be retained, even if making losses, until either they became profitable or could be shown to be dispensable.

The GNRB and other problems, 1958-62

In 1953 the financial position of the GNR forced the two governments to act to ensure that train services on its network would continue. The Northern and Southern Irish governments set up the Great Northern Railway Board, on which the UTA was represented. This arrangement lasted for five years.

By 1958 the GNR financial situation had deteriorated to the point where it was felt necessary, by both governments, to split the GNRB's assets between the two state-owned transport companies, the UTA in the north and CIÉ in the south. From this time the UTA had a legal monopoly of almost all commercial land transport, road and rail, passenger and freight in the province, with the exception of the bus services of the Belfast Corporation Transport Department.

In 1960 passenger services on the Coleraine to Portrush line were withdrawn for the winter, the line opening again only during the Easter and summer holiday seasons. Freight traffic had already been totally withdrawn from this line. In the same year, the Authority withdrew passenger services from the Antrim-Knockmore Junction branch. In Belfast the freight only line between East Bridge Street Junction and Donegall Quay was closed in 1962, to allow for the construction of the Queen Elizabeth Bridge over the river Lagan.

In 1962 the UTA general manager, JA Clarke, set up a committee chaired by the Authority's secretary, JGT Anderson, to examine the financial position of the bus services. Net revenue from the buses had

Erne Bus Services Royal Tiger and Commer buses, with a UTA Tiger Cub behind, at the Custom House, Dublin. This small bus company was taken over by the UTA in 1958.

N C Simmons

fallen from £246,026 to £45,172 in five years and, if the revenue from parcels and private hire work was taken out, the stage carriage services themselves had made a net loss of £119,572 over the period.

It was noted that many UTA services were operated in thinly populated areas which, in Britain, were usually left to independent bus companies to serve. An additional problem for the Authority was the continuing growth in the number of private cars.

The Clarke Committee made a number of suggestions regarding making economies, but its most controversial suggestion was that bus services in counties Fermanagh, Tyrone and Londonderry should be handed over to independents "who are prepared to operate to a lower standard such services as they can make pay".

However, as will be seen, the Northern Ireland government, as a result of the findings of the Benson report into the future of the province's railways, decided to implement a more radical solution and to end the consolidated organisation of public transport in the province.

Some of the pressure that the UTA's management

was now under, was reflected in the fact that in 1959 the Authority published a booklet entitled *Aspects of Relations between the Public and the Authority*, aimed at the general public. This attempted to rebut many of the criticisms being made against it with regard to incompetence, a dictatorial management style, extravagance, bankruptcy, being a financial burden on the government and being indifferent to the wishes of the public. It also stressed the financial damage done to the UTA by the requirement placed on it by the government, to absorb the northern portion of the GNRB.

Basically, the commercial results of the merger of road and rail operations had not been a success. Profits from the bus operations had not been sufficient to off-set the losses from the rail and freight services. As noted above, the Authority had found the absorption of the northern network of the GNR particularly difficult.

By 1960, the fact that the storm clouds were gathering for the UTA was clear from the tone of the introduction to its 12th Annual Report. In this report, during George Howden's chairmanship, the

UTA joint road-rail station at Ballymena. A UTA MED railcar set can be seen in the distance. UTA buses include PS1/2 Tigers, front entrance Tiger Cubs and Leyland Titan PD1/2s with low bridge bodies.

Author's collection

complaint was made that, under the Transport Act (NI) 1958, by which the GNRB was merged with the UTA, a requirement was placed on the Authority that, not later than 1964, it would be in a position to meet all its financial obligations. The Authority had accepted this on the understanding that the inflated capital representing the railway assets would be written off, as would the large sums advanced to the GNRB to cover its revenue losses.

The point was made that, up to September 1960, this promise had not been implemented and had resulted in additional loan charges to the Authority of £632,000, relating exclusively to the GNR capital and advances to cover revenue losses. On top of this, it was claimed, the Authority had incurred, during the two years, an operating loss of about £320,000 on the ex-GNR portion of its rail network.

The report then went on to chastise and then exhort the government as follows:

> The Government is, of course, fully aware of the position as it exists in these respects, but it is no comfort or satisfaction to the Authority that circumstances, so far, appear to have precluded

fulfilment of the original intentions. The Authority can but urge the Government to deal with the situation as quickly and as fully as possible. In view of the long delay which has taken place, the Authority must emphasise the need, indeed the indisputable equity, of making the adjustments retrospective to 1st October 1958.

As well as the problems with the absorption of the GNR, the UTA report identified three further underlying problems contributing to the steady worsening of its financial position:

- The steady growth in the number of motor vehicles, which had tripled between 1948 and 1960.
- The effect of increased labour costs resulting from the implementation of national agreements.
- The re-appearance of inflation and its effect on the prices of materials and consumables.

Again there are some harsh words in the report with regard to these matters:

> In every aspect of its various activities, the Authority has had to contend with, and is working

against, intense and ever-growing competition. It is doing so in the face of much popular prejudice and misconception, while receiving little encouragement in several quarters where a better understanding of the position could not unreasonably be looked for.

The Authority's pleas must have had some effect for, in 1962, the Northern Ireland government, under the Transport (Finance) Act 1962, discharged part of the loan capital liabilities of the UTA, a sum amounting to £11.8 million, plus the aggregate revenue deficiency to 30 September 1961, amounting to £4.1 million.

The £11.8 million written off comprised a division of the Authority's loan capital into £5.3 million relating to the UTA (excluding the GNR section), and the remaining £6.5 million relating to the GNR section. Liabilities remaining totalled £3 million of loan capital from the Ministry of Finance, secured on the undertaking and revenues of the Authority. The Act also authorised the Authority to borrow up to £2.5 million to meet continued losses and certain approved capital expenditure.

However, the financial outlook for the UTA was still not good. In spite of the 1962 financial re-organisation, the cuts in the railway network and improved efficiencies in the provision of bus services, the Authority continued to incur heavy losses, the main blame for which continued to be laid at the door of the railway services.

The Benson Report, 1963

As noted above, in 1962 the Northern Ireland government decided to have another investigation into the position of the railways and appointed an eminent London accountant, Mr Henry Benson, to carry it out.

The Benson committee reported to the Northern Ireland Parliament in 1963 and, although its recommendations were drastic, it did not go as far as to recommend complete closure of what was left of the railway system. It did, however, recommend the closure of the ex-GNR Derry to Portadown main line serving the west of the province and the main ex-LMS (NCC) main line to Derry from Bleach Green just outside Belfast. Although both lines served the city of Derry, they arrived there by totally different routes and this recommendation, if implemented, would have meant that the province's second city

would have been left without any rail connections whatsoever.

Benson also recommended the closure of the Goraghwood to Newry, Newry to Warrenpoint, Dungannon to Coalisland, Knockmore Junction to Antrim, and Coleraine to Portrush branches. All freight traffic originating in Northern Ireland, though not parcel traffic, was also to be removed from the railways. What would be left were passenger services on the main Dublin line from Belfast, the line connecting Belfast with the port of Larne, and the Bangor commuter line.

The Benson Report was followed by a government review of all aspects of inland transport. Benson's recommendations had led to the conclusion, in influential quarters, that a single transport operator, such as the UTA, might not be the best way to organise the province's transport provision. A re-organisation policy was announced which involved:

• A planned run-down of the railways.
• The introduction of a licensing system for road freight to allow private hauliers to provide freight services in Northern Ireland.
• The establishment of a Northern Ireland Transport Commission to take over the assets of the Authority and to operate the railways, workshops and hotels sections of the UTA.
• The establishment of publicly owned undertakings to provide road passenger and road freight services.

The new policy was announced by the Minister of Development, in February 1964, and it amounted to a complete separation of the different branches of transport from each other. The following year, legislation was introduced for the establishment of the new public undertakings.

Between the financial re-organisation of 1962 and the end of the 1964 financial year the Authority's aggregate revenue balance had risen to a £1.35 million deficit. The bulk of this accumulated loss comprised a railway deficit of £1.35 million, an omnibus deficit of £150,000, partly offset by an accumulated road freight surplus of £210,000 and a hotels surplus of £78,000.

Nevertheless, the Authority's 1964 Annual Report announced a proposed additional expenditure of £750,000 on modernisation of the rail network, the bulk of which was earmarked for the purchase of more diesel trains. It was also announced that agreement had been reached to extend one-man-bus operations to ordinary service buses in the hope of achieving additional operating economies.

The second cull of the railways, 1965

In 1964 the UTA was authorised to withdraw certain rail services and all rail freight services. This was implemented in 1965 when the Goraghwood-Warrenpoint branch and the Portadown-Derry main line (the 'Derry Road') were closed, the latter in spite of vigorous local opposition at all levels and right across the community. In the event, this was the last line in the province to be permanently closed, the rest of Benson's recommended closures being set aside.

Later that year, the connection between the Bangor line and the rest of the railway system was removed by the closure of the Belfast Central Railway. The link was physically severed in August by the demolition of Middlepath Street bridge in Belfast. This was done to make way for a road widening scheme, connected to the building of the new Queen Elizabeth bridge over the River Lagan.

A large section of the ex-GNR terminus in Great Victoria Street, including the main Dublin arrival platform, was converted into a new bus station in 1965, although a large proportion of the conversion costs were charged to the railway division's accounts! Passenger handling facilities, a bus park, office and canteen provision and an engineering maintenance area were constructed on this land, mainly taken from the railway sidings. However, this conversion did provide an integrated road, rail and air terminal, as buses to Belfast airport also left from there.

The 1966 and 1967 Transport Acts

In 1965, with the ending of rail freight, Northern Ireland Carriers was set up jointly by the UTA and the Transport Holding Company of Great Britain, to take over the road freight operations of the UTA. New Transport Acts were passed in 1966 and 1967 which divided the passenger services of the Authority into two separate private limited companies, Northern Ireland Railways and Ulsterbus. These three semi-independent companies were themselves subsidiaries of the Northern Ireland Transport Holding Company, in which the Authority's assets was vested. The

NITHC was to provide a limited level of co-ordination and manage the property assets of the three new companies.

The purpose of this new policy was to encourage competition between the various transport modes and to place the profit motive at the centre of public transport management. In 1973 Ulsterbus took over the bus services operated by Belfast Corporation, rebranding them as Citybus. Unlike the current arrangements in the Republic of Ireland, the NITHC made no attempt to co-ordinate the services operated by its subsidiaries. In fact, to the man on the upper deck of a Belfast bus, there was no obvious connection at all between Ulsterbus/Citybus and NIR. To the ordinary citizen, the NITHC was, and is, totally invisible.

The new policy also sought to encourage other private sector transport companies to take over bus stage carriage services which the UTA had found to be uneconomic. Only two such companies were set up – Coastal Bus Services in Portrush and Sureline in Lurgan. Both were later acquired by Ulsterbus. Coastal lasted only until 1974 but Sureline remained independent until 1987.

The profitable UTA hotels were sold to the Grand Metropolitan Hotel group, together with the Belfast city centre site upon which the Europa Hotel was built. This had been the site of Great Victoria Street railway station, and the demolition of much of the station's fine 19th century buildings was undertaken while the station was still in use.

The Duncrue Street workshop facilities were split between Ulsterbus and Northern Ireland Carriers. After 1972 NIR's main workshops outside York Road station were physically separated from the rest of the Duncrue Street complex, by the building of the new M2 motorway. Separate NIR workshops had to be maintained outside Queen's Quay station to service the Bangor line stock.

Review of the UTA years

A major problem resulting from the establishment of the UTA out of the previously independent transport concerns, and never fully resolved during its 19 years of existence, was that of a top-heavy management structure. This suffered from severely divided loyalties between road and rail as some of its senior managers had come from the railway side.

Ex-UTA buses awaiting scrapping shortly after the formation of Ulsterbus in 1967.   David Lawrence

The Authority showed a marked preference, some say bias, towards road operations, which resulted in much of the province's rail network being closed while under UTA management. However, given the basis for calculating the financial efficiencies of the two transport modes, this preference was not surprising. As in the Irish Republic, the vexed question of the differences in operating costs of road and rail service providers was never properly addressed, thus placing the costs of railway operations at a disadvantage to those of the providers of road services.

There was also a clear anti-rail bias in the higher reaches of the Northern Ireland government, which exceeded even that in the UTA's senior management. It was this, probably more than the attitude of the Authority, that allowed the swingeing cuts in the rail network to take place. Significant sums from the British Exchequer for lavish road building and improvement programmes also allowed the Stormont government to be much more cavalier about eliminating a major transport provision, which one senior Northern Ireland government minister described as being as "obsolete as the stagecoach".

That this attitude did exist, is reflected in a bizarre episode which took place at the very end of the UTA's life. In 1966-68, before the establishment of Northern Ireland Railways, the railways traded under the name of Ulster Transport Railways. A Mr John Coulthard had been appointed in January 1966 as managing director of UTR on a five-year contract. Mr Coulthard, a native of Carlisle, was a professional railwayman, with significant experience with British Railways, the British Transport Commission and the British Railways Board.

On 8 May 1966, after a six hour board meeting of the UTA, Mr Coulthard was asked to resign. He refused and so was notified by Sir Arthur Algeo, chairman of the UTA, that the board had decided to dismiss him.

Mr Coulthard claimed that he had been wrongfully dismissed. He indicated that there had been a number of disagreements between himself and the board of the UTR, coming to a head when he granted a five day week to certain grades. His board held that this was outside his authority and had not their approval. Coulthard claimed that he had acted correctly and kept them informed at all stages of the negotiations, but that this was not the real reason for his dismissal. He accused his board of presenting a series of allegations, in writing, against him to the board of the UTA which, in spite of Mr Algeo's statement to the contrary, they had taken into consideration in deciding on his dismissal.

Mr Coulthard indicated that in his opinion, because of UTA policy, a situation had been created in Northern Ireland which would have made it difficult to ensure any future for the railways in the province. He noted that the UTA compensation scheme, which cost the taxpayer about £5 million, had deprived the railways of experienced men, and that there had been an unreasonable delay in bringing in legislation for the new railway organisation.

When the news broke of Mr Coulthard's dismissal on 12 May, it sparked off a two hour lightening strike by railwaymen, who held their managing director in high regard. Mr Coulthard, and his staff, feared that the affair was another step towards the total closure of the railways in Northern Ireland. However, when the matter was brought up in the Northern Ireland Parliament, the Minister of Development said that there was no government intention to abandon the railway system, and the policy of the government was to "maintain and increase" its efficiency.

Mr Coulthard was replaced by a Mr Hugh Waring, an Ulsterman whose transport career had started at the age of 14, as a porter on the BCDR. In the fullness of time, compensation was paid to John Coulthard for unfair dismissal.

On the buses side of the UTA, by 1967 the government had virtually 'thrown in the towel'. It seems that they were prepared to sell most of the business off to a Scottish bus company. Writing 25 years later, the first managing director of Ulsterbus, Werner Heubeck, recalled the view of the Stormont government of the time:

> The Northern Ireland government just wanted the UTA bus problem off its hands. They had over a thousand buses in the fleet and the company was losing a lot of public money. They made it clear that they would be happy for me to reduce the fleet to around two or three hundred and run just the main services. I don't think much thought had been given to providing bus services throughout the Province. They wanted the new company (Ulsterbus) to take only what it wanted as long as it was made to pay its way.

In the event, Heubeck and his board, chaired by Sydney Catherwood, nephew of HMS Catherwood, turned the loss-making UTA bus operation round into profit in its first year, without decimating the service, and continued to operate at a profit thereafter.

The UTA made a number of mistakes in the policies it followed towards rail transportation. Firstly, it was assumed that, by closing railways, traffic would transfer to the road services. It often didn't, as passengers increasingly elected to buy cars and freight customers to buy and use their own lorries. It also failed to recognise fully that closing unprofitable branch lines reduced the flow of traffic to the more profitable main lines, thus making the problem of under-utilisation of those lines worse.

The Authority also failed to see the potential of the growth of commuter rail traffic, particularly into Belfast, from what were rapidly becoming dormitory towns around the city, and the associated problems of traffic congestion though, to be fair, they were not alone in their failure to foresee this particular problem looming up. The closure of the double-track Belfast to Comber line was especially significant in this respect and its absence is still keenly felt by unfortunate commuters on that side of Belfast.

It also failed to foresee the potential of containerised freight traffic, originating or exiting from the province's ports, moving on the Irish rail network as a whole. It removed rail access to the ports of Belfast, Derry and Warrenpoint, all of which have since developed as major freight terminals. The only port to retain its rail access, Larne, was left with only railway facilities for handling passenger traffic from the cross-channel ferries.

It is interesting to note in regard to the above points that the possibility of re-establishing rail links into the Port of Belfast is being considered and there are demands for the re-establishment of the rail links to Armagh, Newry and Dungannon.

The birth of 'Translink', 1995

In 1995, as part of the new approach of government towards the provision of public transport in Northern Ireland, Ulsterbus, Citybus and Northern Ireland Railways were brought under a common board of management. The brand name, 'Translink' was adopted for the new integrated public transport service to be offered by these companies. The

NITHC, however, remains in overall control.

Citybus Ltd had been formed out of the assets of the Belfast Corporation Transport Department as part of a local government re-organisation in 1973. Although technically a separate subsidiary of the NITHC, for all practical purposes it has shared a common top management with Ulsterbus Ltd and the two companies' activities and development have been extensively co-ordinated through this structure.

It is interesting to note some of the old familiar aspects of the UTA returning with Translink. Firstly, the colour green has re-appeared. Translink initially adopted the corporate colour green, which then appeared as the 'theme' colour on all its publications and publicity material. It is also the colour of the new Translink 'T' logo which has replaced the Ulsterbus/Citybus 'paper clip' logo, introduced in 1967, and the NIR logo.

In late 1998 'Translink green' appeared, in the form a green waist-level stripe, on the diesel railcar sets and, in the summer of 1999, on some Citybuses. The new Citybus livery is a two-tone green; 'mint green' and 'Translink green' with a red flash. The Ulsterbus livery is similar, except with a blue flash.

Secondly, in a Translink booklet, *Moving Forward on Public Transport*, published in 1997, we see on page 21 as one of Translink's commitments: "Further development of *bus/rail co-ordination*"(author's italics). Truly a phrase redolent of both the NIRTB and UTA!

Thirdly, in the same publication, same page, a commitment: "to develop [an] integrated fare system ". In transport, as in many other things in life, if one waits long enough, most things come round again – flared trousers, crew cuts, and integrated fare systems.

Finally, in the furtherance of the government's new model transport policy, Translink has identified a number of locations for the development of integrated road/rail terminals. The first of these, the Europa Bus Centre/Great Victoria Street railway station in Belfast, exists as a prototype, although it predates the establishment of Translink and was, in effect, two separate schemes which were not originally planned as an integrated concept.

Two towns which are to get integrated terminals are Bangor and Portadown, the latter being earmarked for development as a key transport node with a road/rail interchange, with the addition of a major Park 'n' Ride facility. Since UTA days, Bangor's bus and railway stations have been on the same site, though since 1967 not as an integrated facility. Work on Bangor's new integrated station began in June 1999 and will be completed in the autumn of 2000. (In one public notice, Magherafelt was mentioned as a candidate for the integrated terminal treatment, which was interesting, as the railway to that town had been closed to passengers in 1950 and the track lifted after 1959!)

Bus/train timetables, showing the road/rail connections, have also made a re-appearance with new timetables showing all Translink services. As

Translink's Europa bus/rail station in Belfast, in July 1999, with the Enterprise disgorging its load.

Author

well as the first comprehensive road-rail timetable published since 1967, area timetables for both Ulsterbus and Citybus, showing road and rail services, made an appearance in 1999.

Specially branded 'Linkline' services have also been set up, with timetables showing road-rail connections. For example, the Portadown 'Linkline' bus service connects the railway station with Armagh, Dungannon, Tandragee and Gilford. Park-'n-Ride facilities are being planned for other provincial road and rail stations.

On page five of Translink's *Moving Forward on Public Transport* booklet it states:

> Already 1,700 new bus/rail connections per week have been introduced and wasteful competition eliminated. The bus and rail operations will increasingly be marketed under the new banner to convey the expectation of co-ordination.

In 1999 railway rolling stock numbers were integrated with those of the bus fleet, for computerisation purposes. This was achieved by the addition of 8000 to the existing fleet numbers of both locomotives and railcar sets.

The 'E-Way' study, 1998

However, some students of Irish public transport history are suspicious that another aspect

of the UTA may be evident in Translink's favouring of bus options.

This suspicion has been fuelled by the stated preference of Translink in their 'E-way' study to re-open the old Belfast-Comber rail line, not as a railway, as NIR had been planning before its absorption into the new entity, or as a tramway such as that planned for Dublin, but as a guided bus-way. However, this would extend along only part of the original line, from the Holywood Arches to Dundonald.

The study into the best way to use this part of the old Comber line, including a public consultation process, was initiated in 1998 and was due to report its preferred option in 1999. As of March 2000 this report has not been published, though it clear that Translink's prefered option is the guided bus-way.

Issue 1 of *E-Way Views and News* was published in March 1999 and sent to those who had registered their interest in the project. It stated that 87% of respondents to the E-Way Road Show recorded themselves in favour of a partially segregated bus-based system. This is hardly surprising, as this was the only system properly presented to the public at these exhibitions.

In the section 'Some of your questions answered', it states, in answer to the question "Why don't you

Translink's new Bangor road/rail terminal. Artwork from leaflet.   Translink

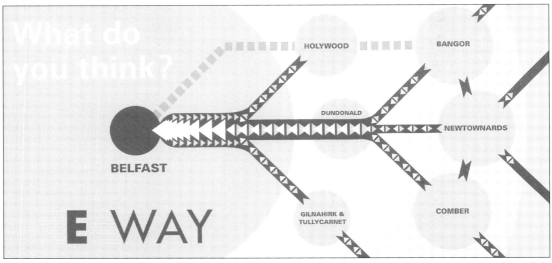

Translink's E-way. Artwork from leaflet.   Translink

put the railway back?" that replacing the railway would now cost an estimated £50 million. It would take much longer to develop and build. The expensive running costs would result in a low frequency service, with few local stops and consequently poor levels of access to the wider community in the corridor. This begs the question, why, if this was the case, Dublin's DART system has been such a resounding success!

There is no mention at all in this publication as to the possibility of an LRT system being an option.

The guided bus-way proposal, if implemented, would channel the E-Way buses into the current road and Lagan bridges bottleneck, which exists between the Holywood Arches and the city centre. This was the very thing that NIR's railway proposals would have avoided.

However, if the 'E-Way' is eventually built as a guided bus-way on the route indicated, it will at least finally secure the part of the old BCDR alignment for public transport use. It will be interesting to see if Translink's expectation of its ability to handle the potentially heavy commuter flows along that corridor will be fulfilled.

The East Belfast guided bus-way would be a pale shadow of the section of the 'Luas' tramway planned for Dublin's Harcourt Street line. Its closure in 1959 was as short-sighted as the UTA's closure in 1950 of the Belfast to Comber line.

Developments at the end of the century

The development of both road and rail public transport services in Northern Ireland looks positive as this book goes to press.

At a strategic level, Translink is lobbying government to be given the same amount of financial support as other bus and rail companies in the UK. Ted Hesketh, managing director of Translink, has stated that, if his company was funded to the same level as transport organisations in Britain, Translink could double the frequency of all its services. The extra passenger revenue so generated would allow the company to reduce its fares by 20%.

As well as the E-Way study mentioned above, other developments are taking place.

Railways

Work on upgrading the cross-border line was completed in 1999 and the new Enterprise trains are now making the Belfast-Dublin journey in under two hours. The main problem for this service is that loadings have exceeded all forecasts and capacity will need to be increased.

Total relaying of the Belfast Central Railway was completed in 1999 and a major refurbishment of

Central Station is planned to follow.

Complete relaying of the Bangor line is also planned and, as noted above, work has begun to re-create a new integrated road-rail terminal at Bangor.

Funding has also been agreed to reinstate the Bleach Green to Antrim stretch of the old NCC main line which, when completed, will knock 30 minutes off the journey time from Derry to Belfast.

As yet however, there are no firm plans to replace the ageing fleet of 80-class diesel railcars with new rolling stock. This is an issue that would need to be addressed as a matter of urgency as the 80-class cars are almost life-expired.

As of March 2000 Translink is awaiting government approval to spend £18 million on six new three-coach train sets. The funding was promised by the DoE in October 1999, but did not materialise. Even if funding is made available now, it would be the end of 2001 before any new trains could be in service. Translink at been in discussion with several manufacturers, including Mitsui of Japan, but at the time of writing no final decisions have been taken.

One of the problems facing Translink is that, unlike train operators in Britain, they cannot lease trains, but must purchase them outright. It is estimated that £100 million would be needed to upgrade the fleet. As a result, the government is examining the possibility of a public/private partnership to fund the purchase of new trains.

Buses

Deliveries of new buses continue apace for both Ulsterbus and Citybus and, with each new type to enter service, the standard of passenger comfort improves. The contrast between the new Volvo B10 low floor buses and the old Bristol RELLs, at least in this respect, is striking. Additional new buses are on order, including a number of additional articulated vehicles (Bendybuses).

A Dennis Trident low-floor double-decker went on trial in Belfast in December 1999. This was the first new double-decker in Belfast since the mid 1970s.

Work continues to replace the rather spartan Ulsterbus depots of the 1970s, with much more attractive bus stations. Examples of the new high quality stations include Laganside (Belfast), Newcastle and Enniskillen.

There are proposals to improve the access of long distance buses into the centre of Belfast during rush hours. The first stage of a bus priority lane along the hard shoulder of the M1, from the Upper Dunmurry Lane junction right into the Europa Bus Centre, has been completed. This will not only be used by existing 'Goldliner' express services, but by a new range of 'City Express' services from the south and west of the city. A City Express service using the M2 to the north of the city was introduced in 1990. It has proved to be very successful, carrying half a million passengers a year, around 25% of them being former car users.

Besides the M1 bus-way, Translink is proposing a huge extension of the current bus lane network. Nine routes into Belfast have been identified, covering 25 miles.

As with the Dublin QBCs, these are designed to speed access to the city for bus passengers and encourage more people to leave their cars at home.

This brings the story of transport policy in Ireland right up to the time of writing, in mid-2000. We now turn to look at the types of train developed by the railways to meet the challenge of road competition and the bus fleets operated by railway companies to exploit the flexibility of the bus to their own advantage.

Chapter 3

Railmotors and steam railcars

About 1850 a number of combined engines and carriages, so-called 'Steam Carriages' were built for light passenger work in various parts of the British Isles. In the north of Ireland such engines were acquired and used by the Londonderry and Enniskillen Railway in 1850-52 and the Londonderry and Coleraine Railway in 1853. The LER acquired one in 1850 and six more in 1852. In the south of Ireland similar vehicles were used about the same period by the Cork and Bandon Railway and by the GSWR.

A LER example, typical of the form, consisted of a small 2-2-0 Adams locomotive, permanently attached to a four-wheeled vehicle combining tender and a single, stage-coach style, passenger compartment. The whole assembly weighed about 11 tons.

The original reason for using such vehicles was that in the early days of railways little passenger traffic existed on some lines, as people had not yet become used to the idea of travelling by train. It was found that a small engine with its own permanently attached stage-coach style carriage, capable of hauling when necessary an additional vehicle or two, was sufficient for the available traffic. However, it was soon found that these early steam carriages tended to be underpowered, too light and too inflexible. On the LER, the Adams patent well-tank engines were seriously underpowered and attracted the unflattering description of 'steam perambulators'.

As the problems of operating this type of vehicle became apparent, the practice of using them, unless for departmental work, ceased until the first few years of the twentieth century, when the first effects of road competition began to make themselves felt. By this time, the name had been changed from 'steam carriage' to either 'railmotors', 'motor trains' or 'steam railcars'.

What was attempted in the early 1900s, by the introduction of railmotors, was a modification of the traditional steam train into a form which was more economical to build and operate. In the opening years of the century, it was hoped that they would be able to deal effectively with the growing threat to passenger traffic which was posed mainly by the emerging electric tramways in the cities. It was also hoped that they would be an economical way to operate poorly trafficked branch lines. Railmotors were introduced around Belfast and Dublin by the GNR to develop commuter traffic, while the single Irish-built GSWR example and the two short-lived DWWR machines were used both on lightly trafficked branch lines and suburban work.

Later, in the 1920s, steam railcars were introduced to try and cope with the growing threat posed by the motor bus in both urban and rural areas. In the north the NCC hoped that using one would stimulate the growth of commuter traffic from the suburbs of Belfast, spreading along their main rail artery into the city. Steam railcars were more popular in the south of Ireland, but were used mainly for branch line work.

One of the attractions offered by several railway companies, were 'house-free' or 'villa' tickets. These tickets were issued to the resident occupiers of newly built houses and were transferable on change of occupancy. They were issued for periods of between seven and ten years. The issue of these tickets induced people to settle in towns and villages alongside the railways, thus creating a greater demand for travelling facilities to and from the city. When the period of validity of the tickets expired, most of the households who had been benefiting from the tickets became regular season ticket holders.

A railmotor or steam railcar consisted of a small steam locomotive permanently attached to a passenger carriage. A duplicate set of driver's controls was fitted into a compartment at the rear of the carriage, which allowed the unit to be driven from either end. This eliminated the associated delays required with normal locomotive practice, as there was no need either to use a turntable to turn the whole vehicle around, or for the locomotive to run around its train at a terminus.

The power units fitted to railmotors were of two basic types. The first type had a vertical boiler, which could be mounted either on a four-wheeled power bogie, located under the front of the carriage body, or on the carriage frame itself (as with GSWR No 1 below). In the latter case, a flexible steam pipe connected the boiler to the cylinders on the bogie. The power unit could be disguised by enclosing it within the carriage body (as with the GNR cars on page 71).

The second type had a miniature horizontal-boilered locomotive of standard configuration, with a normal driver's cab, the carriage portion attached behind this. This second type could have the carriage portion articulated to the engine, or both carriage and power unit could be mounted on a single rigid frame. Railmotors with both types of boiler, rigid and articulated, were used in Ireland.

GSWR railmotor No 1

The GSWR introduced the first Irish railmotor in 1904. It was designed by R Coey, the Chief Mechanical Engineer (CME), and was built in the railway's own works at Inchicore. Coey based his design on that of the two railmotors built by Dugald Drummond for the London and South Western Railway in 1902, from whom he had obtained plans before building his own.

Coey's machine was built for use on the level 5¾ mile Goulds Cross to Cashel branch which had been opened in 1904. The railmotor, which was given the number 1, made five trips in each direction daily. It was not a success and was later transferred to the Drumcondra Link line. This was a poorly patronised suburban service which linked Amiens Street (Connolly) station to the GSWR main line, with intermediate stations at Drumcondra and Glasnevin. It was closed to suburban passenger traffic in 1907. The service was not successful, as it was a roundabout route compared with that offered by the trams.

The railmotor was prone to frequent breakdowns, had an excessive appetite for coal, yet was very low powered, being incapable of hauling a trailer. It was withdrawn in 1912 and the engine portion scrapped at Inchicore in 1915. The carriage section was converted to a normal carriage and became tri-composite brake No 1118. This was scrapped in 1914. It is not clear why it had such a short life as a coach. Incidentally, both of Drummond's 1902-built LSWR railmotors, which had inspired the design of No 1, gave 20 years service, but only after their low powered vertical boilers had been replaced with more powerful conventional boilers.

GSWR No 1 was constructed on a main channel section frame 50'0" long. It had two bogies each with an 8'0" wheelbase. Wheels were 2'9" in diameter. The locomotive section was an 0-2-2T with outside cylinders and comprised a vertical multi-tubular boiler with a heating surface of 393 sq ft and a

Great Southern and Western railmotor No 1 at Inchicore.   National Railway Museum

working pressure of 130 psi. It had outside cylinders 8¾"x 12" and the valves were actuated by Walschaerts valve gear. Drive was to the leading wheels only. It weighed 32½ tons.

The coach portion was divided into three compartments. There was a small first class compartment at the rear, seating six. A short corridor alongside this compartment led to an open saloon seating 48 third class passengers. A luggage compartment was located immediately behind the engine. Access to the coach was from a rear platform which was fitted with fixed steps.

Livery of the locomotive section was black, lined with red and white, while the coach section was dark lake with cream upper panels, lined in gold.

BCDR railmotors and autotrains

In 1904 the management of the Belfast and County Down Railway was having to face the strong possibility of a roadside electric tramway being introduced by Belfast Corporation Tramways (BCT) between Belfast and Holywood. The BCT system had been formed in 1904, mainly out of the assets of the Belfast Street Tramways Company. The BCT had moved quickly to electrify and modernise its network. The proposal for a tram line to Holywood was a revival of a similar proposal in 1897, to introduce horse trams on this route. If the BCT line had been completed, it would have closely paralleled the BCDR's 'water level' route between Belfast and Holywood and presented serious competition for the BCDR's lucrative suburban traffic.

In order to kill the tramway before it got beyond

the planning stage, the directors of the BCDR decided to buy two railmotors to improve the Holywood suburban service. The contracts for horizontal boiler type railmotors were placed in August 1904. Kitson of Leeds built the power units and Metropolitan Railway Carriage and Wagon Co the carriage portions.

After a trial run with press and dignitaries to the railway-owned Slieve Donard hotel in Newcastle, the 'Holywood railmotors', as they came to be called, were introduced, in 1905. Twenty-seven trips were made each day at 20-30 minute intervals. Three new halts were built at Kinnegar, Victoria Park and Ballymacarrett. The railmotors were single class and tickets were issued and collected by conductors, on the tramway principle. So popular were they with passengers, that an order for a third, longer, railmotor was placed in 1907.

The three railmotors were regarded as successful and were used not only on the level Belfast-Holywood route, but soon also on the moderately graded Belfast-Dundonald suburban route. In 1905, the BCDR had found itself in competition, with the Knock extension of the BCT, for the heavy suburban commuter traffic in this area. The railmotors served the suburban stations of Bloomfield, Neill's Hill, Knock and Dundonald on the BCDR main line. However, the engines were worked very hard and No 3 in particular suffered a number of mechanical problems. To try and deal with this, it was swapped with No 1 which took on No 3's heavier coach.

Poor maintenance during World War I left the three railmotors worn out by 1918. At this point the carriages were separated from their locomotives, fitted with a second bogie, and in this form were used with ordinary tank locomotives in an auto-train configuration. The Kitson power units were scrapped in 1924.

During their auto-train phase, the ex-railmotor coaches, which were all-thirds, normally worked with a couple of six-wheeled

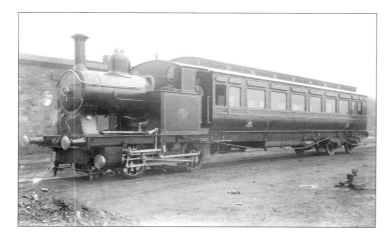

Belfast and County Down Railway railmotor No 1 at Queen's Quay Belfast.
National Railway Museum

coaches, seconds or first/second composites. Motive power for the set was provided by specially fitted Beyer-Peacock 2-4-2Ts Nos 5, 7 and 27. In the auto-train configuration, these engines were capable of accelerating from rest to 60 mph over a very short distance, which made them very suitable for this type of work. Auto-trains were used not only on the Belfast-Holywood service, but some went beyond to Craigavad, serving Marino and Cultra.

The practice of running the carriages in an auto-train formation ended in 1945, as the result of a bad accident, in fog, involving one of the auto-trains at Ballymacarrett Junction. In this incident, just outside the company's Belfast terminus, Queen's Quay, an auto-train ran into the back of a stationary train demolishing several carriages full of shipyard workers, resulting in many casualties and 21 fatalities. Shortly after this accident, one of the GNR auto-trains crashed into the buffers at Great Victoria Street station. This finally sealed the fate of auto-trains in Northern Ireland.

After the Ballymacarrett accident the use of the BCDR auto-trains was discontinued and a GNR railcar was introduced as a replacement. When the UTA was formed, an ex-NCC diesel railcar, No 2, replaced the GNR car on the Holywood turn.

The railmotor carriage portions were given minor modifications and survived in use until the mid 1950s as ordinary open saloons numbered 59, 72 and 173 respectively.

The Kitson power units were small 0-4-0T engines, which were detachable from the carriage portion. These locomotives had 10"x 16" outside cylinders and 3'7" coupled wheels. The boilers were of the normal horizontal type with Belpaire fireboxes. Water was held in a 400 gallon tank located under the carriage body.

The coach sections of Nos 1 and 2 were 50'0" long, over mouldings, 9'2" wide and 12'8½" high over clerestory. They seated 64, all third class, in an open saloon. The larger No 3 was the same height and width but was 60'6" long and seated 80 passengers, again all third class, in an open saloon.

Livery of the coach portion was BCDR crimson lake. The company's coat of arms was carried twice on each side at either end of the vehicle. The number, in figures, was carried above the coats of arms on the waistband. The letters BCDR were carried centrally on the waist band. The number was also carried on an oval plate on the locomotive cab side. The locomotive portion was lined green.

MR (NCC) railmotors

On 1 July 1903 the Belfast and Northern Counties Railway (BNCR) was absorbed by the Midland Railway of England. The new owners arranged to manage their Irish acquisition through a committee, the Northern Counties Committee (NCC).

In 1904 James Cowie, the MR (NCC)'s first manager and secretary, drew up a scheme by which he hoped to develop the outer suburban traffic between Belfast, Greenisland and Antrim, on the NCC main line. He proposed the introduction of a frequent railmotor service along the main line as far as Antrim. To further develop this traffic, new halts were built at Monkstown, Ballyrobert and Muckamore. Railmotors were regarded as especially suitable for this route since at the time the cut-off loop at Greenisland had not yet been built and trains running to Antrim from Belfast had to reverse at Greenisland. The potential delays of this track layout would not be a problem using railmotors.

Two steam railmotors were designed by Bowman Malcolm, the NCC Locomotive Superintendent. They were built at the Midland Railway workshops in Derby and entered service in 1905. Numbered 90 and 91 they were of the horizontal boiler type, consisting of a bogie coach of standard design, articulated to an 0-2-2T locomotive. The carriage portion seated 46 – six in first class, 16 in third class (non-smoking) and 24 in third class (smoking).

The new railmotors were designed for short distances, but were not used as originally intended, the traffic department putting them on the Belfast-Ballymena stopping services, often hauling vans. This required a water stop at Antrim and fast running to keep clear of other trains. As a result, they were continually running hot driving boxes, probably exacerbated by the heavy weight of 16 tons 2 cwt. that the driving axle carried. They were soon worn out and by 1913 were withdrawn. The coach bodies were fitted with new ends and bogies and became brake composites Nos 79 and 80.

Incidentally, the only other railmotors introduced by the Midland Railway, in England, of the enclosed vertical boiler type, were equally unsuccessful.

Midland Railway (NCC) railmotor No 90.

Author's collection

The locomotives had two 9"x 15" outside cylinders, located under the cab. These drove the front pair of wheels only, which had a diameter of 3'7½". The trailing wheels were 3'3" diameter with a wheelbase of 10'0". The total wheelbase, including the coach, was 49'3½". Total length over buffers was 60'5¼". The boiler was 3'8"x 4'6", with a firebox 3'1" long. The inner firebox was 2'2"x 3'2" with 139 tubes 1¼ dia, a heating surface of 313 sq ft and a working pressure of 160 lbs. Tractive effort was 3,798 lbs and the total weight, including coach, was 39 tons 3 cwt. Water was carried in a 500gal. tank under the coach. Coal capacity was 11½ cwt.

Livery was Midland Railway crimson lake on the coach, with the locomotive portion in black. The vehicle number was carried on a plate affixed to the locomotive cab side. The company crest was carried on the coach side. Class designations were carried on the door waist panel in words.

GNR railmotors and push-pull units

The Great Northern Railway (Ireland) obtained three railmotors of the enclosed vertical boiler type in 1905 from the North British Locomotive Co of Glasgow and, in 1906, four from Manning Wardle of Leeds. Coachwork for the North British type was by T Pickering of Wishaw and that for the Manning Wardle vehicles was by Brush Electrical Engineering of Loughborough. The North British units were to be used in the Belfast area on Belfast-Lisburn suburban work and the Manning Wardle units in the Dublin area on the Dublin-Howth suburban service.

However, there is photographic evidence that the Manning Wardle units later also operated in the Belfast area. On both routes there were opportunities for increasing traffic.

In the Belfast area, this increase came from the spread of housing along the Lisburn Road (originally a turnpike road when the railway was first built), which the railway closely paralleled. It also came from the developing town of Lisburn itself and from intervening villages such as Finaghy and Dunmurry. The main competition was from the municipal tramway, which ran about three miles as far as the village of Balmoral. The GNR station was close to the tram terminus. Shortly after its establishment in 1905, the BCT put forward plans to extend its services beyond Balmoral and this spurred the GNR to introduce steam railmotors and, in 1907, to construct additional halts at Finaghy, Derriaghy and Hilden for their use.

In the south of the country, the Howth branch, which was 3½ miles long, left the GNR main line about 4¼ miles north of Dublin and was built to serve the residential areas of Sutton and Howth. All the trains made connections with the GNR Hill of Howth trams at both Sutton and Howth. Stations between Amiens Street (now Connolly) and Howth Junction stations were also sources of suburban traffic.

The main competition for the Howth traffic came from the Clontarf and Hill of Howth Tramway. This line abutted end-on to the Dublin United Tramway Company's line at Clontarf and ran to the east pier at Howth harbour, close to the GNR station. Trams ran right through from O'Connell Street to Howth, although the railway and the tramway ran close together only from Sutton Cross to Howth. The tram route between Malahide Road and Sutton took the coastal route, while the railway ran further inland via Raheny, Kilbarrack and Howth Junction.

The railmotors were fitted with low-level platforms or running boards, which should have

GNR railmotor No 2.
National Railway Museum

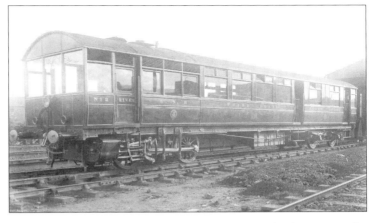

meant that alighting from them would have been easy. This should have meant that normal station platforms would have been unnecessary at the halts specifically constructed for the use of the railmotors. However, the GNR still found it advisable to provide such platforms in order to avoid possible litigation from clumsy passengers. Each railmotor could operate alone or with a trailer coach attached. When running with a trailer coach they ran with the power bogie adjoining the trailing coach. Four special saloon coaches were built to operate as trailers for the railmotors and each trailer coach had a driving compartment fitted to one end. Five similar vehicles were built for push-pull operation.

Once in operation, the railmotors were not regarded as a great success, being uncomfortable in running and giving a lot of trouble in maintenance. They ran only for a short time on the Dublin-Howth services, before being replaced by larger capacity push-pull trains. On the Belfast-Lisburn service they were regarded as rather more successful, apart from the maintenance problems. However, all seven were withdrawn in 1913 and converted into normal, if somewhat longer than standard, coaches. The Dublin coaches became Nos 204-207 and the Belfast ones became Nos 201-203.

The seating capacity in the railmotors was 59, comprising 20 first class and 39 third class. Nearly a million passengers were carried by the railmotors in

GNR autotrain at Irish Street halt, Armagh, in 1909, while working the Keady branch.　　Author's collection

an average year.

The Manning-Wardle power unit had a vertical multi-tubular boiler, a heating surface of 653.1 sq ft, 11½ sq ft of grate, and a working pressure of 175 lbs. It had four coupled wheels of 3'9" diameter, on a wheelbase of 8'0", with outside cylinders 12"x 16" and steam distribution by Walschaerts valve gear. Water was carried in a tank suspended under the floor of the coach. The first of the units was 58' 0" long; the remainder were 61'6".

The North British unit was again of the vertical boilered type with a tube heating surface of 660 sq ft. It had four coupled wheels of 3'7½" diameter on a wheelbase of 8'0", with outside cylinders 12"x 16" and Walschaerts valve gear. Water was again carried in a tank suspended under the floor of the coach. The weight on the motor bogies was 25½ tons and the total weight of the unit 40½ tons.

The GNR replaced their railmotors on this suburban work with auto-trains. Auto-trains were also used on other duties. For example, coach No 10 was used on the Keady-Armagh branch shortly after it opened. An auto- or push-pull train consisted of a standard BT class 4-4-0T locomotive placed between two specially designed coaches, which had driving compartments at each end of the train. The locomotive was controlled by shafts and levers, and a telegraph in the locomotive cab was used to indicate to the fireman the required position of the reversing lever.

These two-coach units could accommodate 124 passengers, 30 first class and 94 third class. If required, the engine could run with a single coach. The BT class tank engines had 4'7½" coupled wheels, 15"x 18" cylinders, and weighed 31 tons 10 cwt.

The push-pull units were regarded as a success by the GNR but, as noted above, were finally withdrawn after two accidents involving such trains – the Ballymacarrett accident on the BCDR and the Great Victoria Street accident involving one of the GNR auto-trains.

The livery was varnished mahogany lined in gold and blue. Numbers and lettering were in gold blocked in blue. The GNR coat of arms was carried twice on each side. Class designations were by word and carried on the waistband and the fleet number at each end, also on the waistband.

DWWR railmotors

The DWWR acquired two railmotors in 1906 from Manning Wardle of Leeds. These had been designed by Richard Cronin, the Locomotive Superintendent, and were of the normal horizontal boiler type. They carried the numbers 1 and 2.

They were built for use on a new service between Bray and Greystones. The first of the two entered service in the month in which it was delivered, operating for a week between Landsdowne Road and Ballsbridge sidings in connection with a show at the Royal Dublin Showgrounds. It was then transferred to the work for which it had been designed.

The railmotors made ten return trips daily, with one intermediate stop at Bray Head halt, taking 13 minutes for the journey. However, they were not successful and had very short operational lives. This was chiefly due to the fact that their riding characteristics created extreme discomfort for passengers, particularly those in first class. As Kevin Murray has wryly stated in his book *Ireland's First Railway*, that was probably the worst defect they could have had! Both cars were withdrawn in 1907 and the locomotive portions were separated from the coach portions in 1907-08.

As originally built, both railmotor engines were 4-coupled with Belpaire boilers, 12"x 6" outside cylinders and 3'7" driving wheels. No 1 was fitted with Walschaerts valve gear and No 2 with Marshall's valve gear. Both were fitted with well and side tanks. Boiler pressure was 160 lbs; grate area 9¾ sq ft; tractive effort 7,287 lbs; weight 26 tons 14 cwt. The driver, when operating from the coach driving compartment, communicated with the fireman on the locomotive footplate via a telegraph similar to those used in ships. However, he did have a brake at the coach end.

After withdrawal as railmotors, the locomotive sections were renumbered 69 and 70 respectively and were used for shunting. No 69 was rebuilt in 1914 as a 2-4-0T and again in 1925 as a 0-4-0T. They were later named by the GSR, No 69 as *Elf* and No 70 as *Imp*. *Imp* was sold to the Dublin and Blessington Tramway but was later returned, and *Elf* finished up in Limerick as station pilot. *Elf* was withdrawn in 1928 and *Imp* in 1925. Both were scrapped in 1931.

The coach portions, which seem to have been built by Brush of Loughborough, were 44'0" long and

DWWR railmotor No 1.   National Railway Museum

weighed 22 tons. They seated 16 first class and 39 third class passengers in open saloons and were articulated to the engine. The first class compartment was at the rear and the third class next to the engine. There was non-smoking accommodation in both first and third class.

After withdrawal, the carriage portions were converted into composite coaches. They were renumbered 19 (GSR 209D), and 20 (GSR 212D) respectively. As coaches 19 and 20 they are recorded as being tri-compo with one third-class, four second-class and two first-class compartments. They remained on the books until 1960.

The livery was lined black for the engine, which carried the legend 'No 1' (or 2) on the centre of the side tanks. Coach livery was reddish brown with gold lining. The legend 'Dublin Wicklow and Wexford Railway' (in capitals) appeared along the waist panels, the full length of the coach. The number appeared again on the coach end nearest the engine, on the waist panel.

By the time of the withdrawal of the railmotors, the DWWR had been re-named the Dublin and South Eastern Railway, but it is unlikely that in their railmotor form they would have carried the DSER livery.

In 1922, under continuing pressure from tramway and growing bus competition, the DSER converted six bogie coaches to what they called 'bus coaches' for use on the Dublin-Greystones route. They had accommodation, smoking and non-smoking, for both first and third class passengers. The coaches were fitted with centre vestibules with sliding doors and it was intended that tickets would be issued by conductors, tramway-style.

In the event, they were not used as intended, being operated with normal six-wheeled carriages carrying luggage and parcels and the conversion probably only served to deprive the company of the use of valuable bogie vehicles.

Steam Railcars, 1925-28

LMS (NCC) Sentinels

On 1 January 1923 the MR (NCC) found itself part of the London Midland and Scottish Railway Co. The new owners decided to keep in place the system of management for their northern Irish subsidiary set up by the Midland Railway, namely the Northern Counties Committee. The manager and secretary of

the NCC at this time was James Pepper, a Derby-trained Englishman, appointed in 1922 by the Midland Railway. On taking up office in Belfast, Pepper decided to institute a broad programme of economies and improvements right across the NCC.

As part of this programme, two steam units were ordered in July 1924 from the Sentinel Waggon (sic)

Works Ltd of Shrewsbury. The production of Sentinel, and the slightly later Clayton, steam railcars represented a second attempt by various railway companies to make something of the railmotor concept after the failures of those introduced during the Edwardian era.

One of the NCC vehicles was a light locomotive for shunting or working branch line trains, the other a steam railcar for local services. The railcar was numbered 401 in the coach series and the locomotive 91, carrying the number of one of the two earlier steam railmotors.

The carriage portion of the railcar was built by Cammell Laird and Co of Nottingham and seated 44. The vehicle was 56'6" overall. In fact, this vehicle was part of a larger group of orders for 13 units placed with Sentinel by the LMS between 1925 and 1927, although the NCC vehicle seems to have been introduced a year ahead of the vehicles in the main LMS order.

Sentinel railcars were designed specifically to counter the increasing competition from road vehicles and Sentinel claimed that its railcars could be operated at less than half of the running costs of a conventional locomotive-hauled branch line train. Significantly, it also claimed to be able to offer lower operating costs per seat-mile compared with a road bus. At the time that railcar 401 was acquired, the NCC built additional halts on the Belfast-Larne line.

These were all in the Carrickfergus area, at Clipperstown, Barn, Downshire Park and Eden, all close to the main Belfast to Larne road. These new halts would have been located specifically to tap into the growing commuter traffic between Carrickfergus and Belfast.

However, when introduced by the NCC in 1925, the Sentinels proved to be badly underpowered, suffering continuously from shortage of steam due to the smallness of their boilers. Also, the driving chains broke too frequently. The gradients over which the Sentinels were required to operate proved to be rather severe for this particular type of unit. As a consequence, the locomotives were regarded as very unsuccessful. This is surprising, considering how successful Sentinel units were in other parts of Britain, particularly on the LNER, and elsewhere in the world.

A possible explanation is that the NCC railcar was an early version of the later successful Sentinels, and was based on the units provided to Jersey Railways. Sentinel had only begun to manufacture these locomotives in 1923, the first examples being sold to Jersey Railways in that year. In fact, an early example supplied to the LNER for trials in August 1924, exhibited similar problems to those experienced by the NCC railcar.

Sentinel vertical boiler units came to be regarded as the most successful of this type of configuration.

LMS (NCC) Sentinel railcar No 401 and shunting locomotive No 91 at York Road, Belfast.

HC Casserley

The company produced far more than any other maker and were only superseded in the market by diesel railcars in the 1930s.

The NCC Sentinels were both withdrawn in 1930 and scrapped in 1932. The body of the railcar ended up doing duty as a scout hall beside Maghera Orange Hall, and survived until the early 1980s.

However, because of increasing road competition the LMS (NCC) stuck with the railcar concept and enthusiastically developed internal combustion engined versions in the 1930s.

Both locomotive units acquired by the NCC were of the vertical boiler type, with boilers 2'8½"x 4'4½", working pressure of 275 lbs. Cylinders were 6¼"x 9" and driving wheels were 2'6". The wheelbase of the locomotive was 8'8". A chain driven transmission from the vertical cylinders was used. Water capacity was 300 gallons and coal capacity was 13 cwt. The carriage section of the railcar was articulated to the locomotive. The whole assembly weighed about 20 tons. The NCC and the later GSR Sentinel railcars were all of the first generation 'lightweight' type without proper draw gear. None of the later heavier and larger Sentinels, such as those used by the LNER, operated in Ireland.

The livery of the railcar was LMS maroon. The vehicle number was carried on the power unit at waist level. The legend 'LMS NCC' was carried half way along the coach portion also at waist level.

In an article entitled 'Recollections of the BCDR', published in Vol 3 of the *Journal of the Irish Railway Record Society*, W Robb relates an interesting anecdote from his boyhood:

> Another early memory is of coming home from school one day and hearing a vehicle with a very unusual sound coming along the line; this turned out to be what I later learned was the NCC Sentinel locomotive having a trial run on the B&CDR.

This is the only reference to the BCDR being interested in Sentinels that I have ever come across.

GSR Sentinels

Four steam railcars and two locomotives supplied by the Sentinel Waggon Works Ltd were introduced by the GSR in late 1927. The railcars were numbered 354 – 357 in the GSR coaching stock series.

One of the railcars was used on Cashel-Goulds Cross branch, while the others ran in the Limerick area on local services. The railcars were fairly successful machines, one, No 356, being still at work on the Foynes branch in 1939. The two locomotives were used as shunters.

Great Southern Railways Sentinel railcar No 354.   National Railway Museum

They all remained on the books until the early 1940s, although probably out of use during the last few years of their lives.

Like the NCC cars, both shunters and railcars were fitted with a four-wheeled, chain-driven, power bogie equipped with a vertical boiler and vertical cylinders 6¾"x 9". The boiler pressure was 275 lbs and the driving wheels were 2'6" in diameter. They were fitted with vacuum brake cylinders on the power units, operating expanding brakes on all four wheels fitted with brake drums. The 400 gallon water feed tank was positioned above the driving bogie and filled through panels on either side of the car. Coal bunkers were located at the front left hand side opposite the controls and had a capacity of 15 cwt. This gave the cars a range of about 120 miles.

On the railcars the totally enclosed power bogie unit was articulated to the carriage portion which accommodated 55 third class passengers.

All these vehicles were built in 1927. Nos 354 and 357 were withdrawn in 1941 and Nos 355 and 356 in 1942. Both shunters, Nos 281 and 282, were withdrawn in 1948.

The Sentinel railcars carried GSR coach livery which was a dark purple-brown with straw lining. The fleet number was carried at waistband level on both the power unit and the carriage portion The GSR arms were carried half way along the carriage portion, just below the waistband.

GSR Clayton steam railcars

Six steam railcars were acquired in 1928 from Clayton Wagons Ltd of Lincoln, at a price of £1,800 each. They were identical to those used on the LNER, except that the GSR cars carried a small number of first class passengers whereas the LNER cars were third class only. They were supposed to be capable of a speed of 45 mph. These were the only Clayton railcars to run in Ireland.

Their numbers were also in the coaching stock series, 358 – 363 following on in sequence from the Sentinels. They were used on the Westland Row (Pearse) to Dalkey service, the Harcourt Street to Foxrock service, the Cork to Macroom line (closed in 1935) and on the Athlone to Sligo line, but were not regarded as a success.

In an IRRS Journal article on these vehicles, RN Clements gives a valuable insight into the practical problems of operating them. He noted that No 360, when in use on the Foxrock service, was very unpopular with the crews. It seems it could not make enough steam to cope with the severe gradients on the line. The boiler tubes were close together and, as the canal water used at Harcourt Street was of poor quality, this required the boiler to be washed out frequently. This problem was compounded by the inaccessibility of the washout plugs.

Clements also noted that there were complaints about the difficulty of firing the vertical boilers. The boiler was fired from the top, through a stoking tube, and firemen complained that this arrangement gave little control over the placing of the coal which just piled up in the middle, unburnt. However, Clements believed that the main problem here was that the firemen hadn't enough experience of firing these cars and were putting on too much coal at one time. Coal consumption on the Harcourt Street line was said to be 19 lbs per mile, and on the Westland Row line, 25 lbs per mile. This contrasted very unfavourably with the makers claim of 10-11 lbs per mile.

When withdrawn in 1931, the power units were removed and the coach portions converted into three two-car articulated sets which then had a further, but more successful, lease of life on the Waterford to Tramore line.

The Clayton cars used essentially the same standard engine and boiler unit as was fitted to the Clayton 'undertype' road wagons. The boiler was of the vertical cylindrical water-tube type and was fitted with a superheater. Working pressure was 300 lbs. As mentioned above, it was fired through the top via a stoking tube.

The engine was of the non-compound fully-enclosed type, with gear transmission. It was a two cylinder, horizontal, high-pressure, totally enclosed unit. The cylinders, crank-case and the gearbox for the reduction gearing were bolted together so as to form one unit. There were two double-acting cylinders 7"x 10" to which steam was distributed by piston valves beneath the cylinders. The valves were operated by eccentrics mounted on a layshaft driven by spur gearing from the crankshaft. The cut-off was variable in forward and reverse between 80% and 30%.

The engine drove the wheels by a crankshaft pinion in connection with a spur wheel on the axle.

GSR Clayton railcar No 362 at Glanmire Road, Cork.   Author's collection

The two axles of the driving bogie were connected by outside coupling rods. Bogie wheelbase was 7'0"; driving wheels were 3'6" in diameter. The boiler assembly was located over the rear of the driving bogie, within the front part of the car body.

The body and underframe were built as a unit with steel framing and exterior panelling. Unlike the Sentinels, where the locomotive portion was articulated to the passenger coach, a Clayton cars body was mounted on two four-wheeled bogies, one of which carried the power unit. Swing-bolster bogies were fitted, with laminated side-bearing springs and helical bolster springs. Clayton cars were, in effect, rigid railcars with pivoting power bogies

The water tank and coal bunker were mounted forward of the engine bogie, outside the car body. The water capacity was 550 gallons and coal capacity was 15 cwt. Total weight of the railcar was 25 tons. They could be driven from either end. Seating was nine first class and 55 third class.

They carried same livery as the Sentinels. The fleet number was carried on the side, near the front of the coach body, with the company crest repeated twice on each side. The class designations '1' and '3' were carried in large figures on the doors.

The CVR Sentinel railcar proposal

In spite of having introduced a very successful Walker's diesel railcar in 1932, in the late 1930s the Clogher Valley Railway proposed the purchase of a Sentinel steam railcar. This was to be a further attempt to reduce operating costs, in a last ditch attempt to save the railway from closure.

The decision to examine this option was taken by the company, in concert with Fermanagh and Tyrone County Councils, through whose areas the railway ran. These two councils had controlled the railway since 1928 through a joint management committee.

Sentinel Waggon Works Ltd were contacted and supplied photographs and drawings for what they considered a suitable vehicle. However, the idea came to naught and the CVR was closed in 1942.

Chapter 4

Petrol, diesel and battery railcars

The term 'railcar' covers a broad spectrum of vehicle types, from early car or bus-like vehicles to later large, sometimes multi-engined, full-sized railway vehicles. Depending on the company which used them, they were called 'railmotors', 'railbuses', 'railcars' or 'autocars'.

There had been a number of early attempts, mainly in England, to apply the petrol internal combustion engine to rail traction, but these were mostly failures. The real forerunner of the modern diesel train appeared in the 1920s with the development of the petrol and later, more importantly, the diesel railcar. In Ireland these developments occurred about the same time in a number of companies. In some cases, very successful vehicles, from both mechanical and operating points of view, were produced.

Many railway operators were attracted, for economic reasons, to the use of diesel traction as an alternative to steam. Steam locomotives have a low thermal efficiency, around 5-6%. At a time of high coal costs, employing diesel traction, with its higher thermal efficiency of 25-35%, was most attractive.

In addition, diesel traction with its variable overall gear ratios, provided faster acceleration from rest than steam could provide. This was particularly important for trains, such as those used on commuter services, which had to make frequent stops.

There was also the attraction of standardisation of parts with those of buses. This was particularly important for companies like the LMS (NCC), GNR and later UTA and CIÉ, all of which had considerable bus fleets.

In this chapter, it is intended to concentrate on the evolution, from small beginnings, of the vehicles which led eventually to the development of the multiple-unit trains of the 1940s and 1950s. The vehicles covered have in common the fact that they were not conversions from road vehicles, but were built exclusively for railway use to meet the growing challenge from road competition.

Most Irish railcars used bus engineering technology. They were self-contained units, although it is recognised that in some cases they were capable of hauling a trailer vehicle. Although some may have operated in pairs, they were not capable of operating in multiple, one unit either being hauled 'dead', or each unit requiring a separate driver.

In addition, five examples are included of vehicles which, although built as self contained railcars, were capable of working in multiple with other similar vehicles which were designed to operate purely as multiple units. Three of the five examples come from the UTA's Multi-Purpose Diesel railcar programme of 1961-2 and the other two from IÉ's 2700 series 'Arrow' railcar fleet of 1999. These five vehicles, separated in time by nearly 40 years, are the direct descendants of the GNR and NCC single-unit railcars, and were designed to handle much the same sort of traffic.

One oddity covered in this chapter is the solitary NIR BREL-Leyland National railbus, which was introduced in the 1980s. It did not influence multiple unit train development in Ireland, although it did in Britain. However, it meets all the other criteria for inclusion in this chapter.

The chapter will be divided into three sections; the first will list pioneer vehicles; the second will deal with the mainstream vehicles; and the final section will cover those projects which did not develop into anything permanent, the 'might have beens' of railcar development in Ireland.

Pioneer vehicles 1907-1925

CDR

Road competition in County Donegal developed after World War I as the result of an influx of surplus, cheap ex-military road vehicles, and the Co Donegal Railways decided on a strategy of modifying their railway operations. Additional flexibility was introduced by allowing trains to pick up and set down almost anywhere in response to customer demand. Henry Forbes, the CDRJC's energetic manager, was an instinctive marketing man! In order to make these changes possible the CDRJC decided to experiment with the use of internal combustion engined railcars.

The company had pioneered internal combustion using a four wheeled vehicle which it had bought in 1907. This little petrol engined vehicle, built in Birmingham by Allday and Onions, originally had an open body and was used as an inspection car. In 1920 it was fitted with a closed body and was re-engined with a more powerful Ford 22 hp engine. In this form, it was about the size of a contemporary light road van and could carry six people. The wooden body was 8'6" long and 3'7" wide, with windows in the sides and rear like a bus. This little vehicle was used variously as an inspection vehicle, for weed-killing duties and for carrying mail but, importantly, during the strike of 1926, it was also used for carrying passengers. After being re-engined, it was capable of hauling a lightweight trailer which seated an additional 25 passengers. In 1949 it was again re-engined with a 36 hp Ford engine.

Although used only occasionally for passengers, it yielded valuable information on the potential of this type of vehicle. It pointed the way to operating a passenger service at lower cost than steam traction could provide. It also paved the way for a number of railcars of various types and sizes, culminating in the very successful Walker railcars introduced in the 1930s.

MGWR

The MGWR introduced a small petrol railcar in 1911, for use on the lightly trafficked Achill branch. It had the advantage that it could stop at level crossings and could therefore be used to collect both passengers and mail. However, it was not a great success, and in 1916 was transferred to the Civil Engineer's department as an inspection car, with the running number '1'. It was withdrawn in 1933.

It was built by Charles Price and Sons of Manchester. It had a four cylinder, 27 bhp petrol engine which could deliver a maximum speed of 53mph. This was replaced in 1914 with a new engine. Interestingly, the car could be driven from either end which precluded the need to turn it at termini. It was fitted with reversible seats and could carry 11 passengers. The body had lockers, which projected like car boots, at either end of the body for the carriage of luggage and mail.

County Donegal Railway Ford railcar No 1 and trailer No 5 at Stranorlar.   National Railway Museum

TDR

In 1922 the Tralee and Dingle Railways built a little four-wheeled railcar, not unlike CDR No 1. It was constructed on a Baguley chassis and was equipped with a Ford Model T engine and transmission. It had a four-seater four-door body and was used for line inspections. It may also have been used for military scouting in its early years.

It was moved to the West Clare by the GSR on the amalgamation in 1925, but returned to the Tralee and Dingle in 1953 in connection with a proposal to operate the line with railcars, similar to those used on the West Clare. However, nothing came of this idea.

Though finally stationed at Ennis, it could be moved for inspections to other parts of the GSR narrow gauge system on a standard gauge low loader. It was used also on both the Cavan and Leitrim and Cork narrow gauge lines until their closure. Following closure of the West Clare system, it was scrapped at Ennis in 1961.

The car was 10'6" long and 6'0" wide. Wheels were 2'0" in diameter. It was re-engined in 1938 in Ennis, with a more modern Ford 8 engine replacing the original Ford Model T engine. Engine and transmission were mounted on a steel frame which carried the body.

The original livery was probably the Tralee and Dingle's dark green. Later, it was numbered 6C in the GSR Civil Engineer's inspection car series (see page 83). It was then in GSR red with the crest on the rear panel and the number on the front door. The crest and number were later dispensed with. The final livery was CIÉ green.

CVBT

In 1925 the Castlederg and Victoria Bridge Tramway in Co Tyrone built, in their own

Midland Great Western railcar. Introduced 1911. Withdrawn 1933.
Author's collection

Tralee and Dingle inspection car at Ennis in 1961, after the closure of the West Clare system. Colour-Rail NG91

workshops, a railcar powered by a 20 hp Fordson TVO paraffin tractor engine. It was a four-wheeled vehicle driving on all four wheels through a chain drive. It could be driven from either end.

In its original form, it was under-powered and although more economical than steam traction, it was not regarded as a success. It lasted for three years before being withdrawn.

In 1932 the CDR bought it for £25. It was in poor condition, abandoned behind the CVBT running shed. Its engine had been removed and sold. It was brought to Stranorlar where it was fitted with a 22 hp Reo FA petrol engine and a new body. It became the second CDR No 2, entered service in 1934 and ran in this form until 1941.

In 1944 the engine was removed, another body was fitted and the frame was lengthened by three feet. In this form, it ran as trailer No 2 until the closure of the system.

Petrol railcars 1926-1931

County Donegal Railway

By 1926, with road competition building up and with the need to reduce operating costs now paramount, Henry Forbes, general manager of the CDR, decided to show that the railway could both increase flexibility and reduce operating costs, by using railcars.

The Derwent Valley Light Railway in England was disposing of two Ford petrol-engined railcars. The CDR bought them, had them re-gauged in the GNR Dundalk works from 4'8½" to 3'0" and had the bodies lowered to match the level of the CDR station platforms. As rebuilt, they were 5'10" wide by 7'3"long. On the DVLR they were run back-to-back, but on the CDR they ran separately, and were one-man-operated. They were numbered 2 and 3 and were used on regular passenger services, setting down and picking up passengers almost anywhere.

Operating costs were about a quarter of those of conventional steam trains. They could each carry 17 passengers which was sufficient for many of the services on which they were used. The CDR used these vehicles until 1934, when they were withdrawn and scrapped.

In 1927 the Londonderry and Lough Swilly Railway Company approached the CDR with a view to the latter taking it over. The Joint Committee toured the whole of the LLSR network in Railcar No 3. In the event, in spite of a request from the Northern Ireland government, supporting the Lough Swilly's proposal, the CDR decided not to agree to this take-over. Thus, No 3 became the only example of a railcar operating over Lough Swilly metals.

In 1928 railcar No 4 was built for the CDRJC at the GNR's Dundalk works. It was again a Ford petrol engined vehicle, utilising a Ford lorry chassis, with bodywork built by the Strabane coach builder O'Doherty. It could carry 20 passengers.

No 4 was sent on loan to the Clogher Valley Railway for trials, after which the CVR introduced its own railcar. No 4 gave 19 years service to the CDRJC and was scrapped in 1947.

In 1931 CDR No 6 appeared (No 5 being a trailer) from Dundalk, where it had been designed, again the bus-type bodywork being provided at Strabane. It was a heavy vehicle for its size, at 5½ tons and was powered by a second-hand Reo 32 hp petrol engine.

It incorporated a number of features which had been learned from problems with the earlier railcars; for example, it was fitted with solid railway-type axles rather than road axles to eliminate the problem of axle breakages. It had no differential and this, plus the provision of a front pony truck, allowed it to operate over trackwork very smoothly. The two rear axles were coupled together by a chain as a method of eliminating the skidding on ice which had plagued the earlier railcars.

It ran until 1945 when it was converted into a four-wheeled trailer and survived in this form until 1958. It was to be the last of the petrol engined railcars.

The first County Donegal Railway No 2. This vehicle was one of a pair bought from the Derwent Valley Light Railway.
National Railway Museum

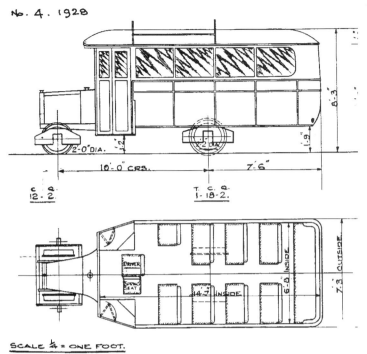

No. 4. 1928

SCALE ¼ = ONE FOOT.

CDR railcar that could be driven from either end, though for some reason this did not influence the design of later purpose-built railcars supplied to the CDR. It had its engine removed in 1943 and was rebuilt as a trailer.

CDR railcar trailer No 13 also started life as a railcar on the Dublin to Blessington tramway. It was one of a pair of Ford railcars acquired in 1925 by the DBST. These two cars were built on Ford Model T chassis with Ford 22hp petrol engines. Bodywork was constructed by the DBST using parts of two petrol-electric trams the company had unsuccessfully introduced in 1915. Numbered 1 and 2, the two railcars could each accommodate 16 passengers and were used for off peak services as a way of reducing operating costs.

On the closure of the Dublin and Blessington line, the CDR bought one of these cars, removed the engine, had it re-gauged and used it as railcar trailer until 1944, when it was scrapped.

DBST railcars

A single-decker, double-ended railcar had been bought in the mid 1920s by the Dublin and Blessington Steam Tramway, in an attempt to reduce operating costs. In 1934, on closure of the Dublin and Blessington, this railcar was bought by the CDR.

This Drewery built vehicle had 35 hp petrol engine, and seated 40 passengers. It had two driving axles, as well as two trailing pony trucks and was thus eight-wheeled. The ends were vestibuled. However, the motors, encased in a wooden protective cover, protruded through the floor of the passenger saloon and were an obstruction.

As originally built, it was a standard (5'3") gauge vehicle. In Stranorlar it was re-gauged to 3'0". It became CDR No 3, the same as its DBST number, and was the only

CDR railcar trailer No 3, ex-Dublin and Blessington railcar No 3, at Stranorlar.   N Simmons

GSR Drewry railcars

In 1926-27 the GSR authorised the purchase of eight petrol driven four-wheeled railcars. These comprised four inspection cars (Nos 2-5, Works Nos 1492-5, built in 1926-27) for use by their Civil Engineers; two passenger-carrying railcars for the standard gauge system (Nos 385 and 386 in the GSR coaching list, built in 1927) and two for use on the narrow gauge West Clare section (Nos 395 and 396, also 1927). They were ordered from the Drewry Car Co Ltd, but were actually built by EE Bagueley Ltd of Burton on Trent, through a working agreement between the two companies.

The narrow gauge cars were 20'6" long by 6'9" wide with four bays of windows. They weighed 5 tons 10 cwt in working order. The wheels were 2'0" in diameter and the wheelbase was 8'9".

The standard gauge cars were about 3'6" feet longer, with five window bays. Their wheelbase was also longer and the wheels were of a slightly larger diameter than those of the narrow gauge cars.

Each car was fitted with a 40-45 hp Bagueley four cylinder petrol engine. This engine was rated at 40 bhp @ 1000 rpm, 46 bhp @ 1050 rpm. Cylinders were 4¾"x 6" and the engine was mounted longitudinally at one end of the car. They were fitted with both electric and hand starters. Each car had a 15 gallon fuel tank and a 12 gallon water tank for engine cooling. Transmission was by a cone clutch and a Bagueley three-speed gearbox and the final drive was via chains to the rear axle. They had a top speed of 40 mph. It was estimated that they had running costs of about 9d per mile.

Controls were provided at each end, so they did not need turning. They were not fitted with vacuum brake connections. They had standard couplers appropriate to the gauge the cars were built for. When new, the narrow gauge cars were fitted with large electric headlamps, but these were later removed.

The bodies were carried on steel channel frames, which had an inner frame on which the engine was supported. The interior was panelled with plywood to waist level and all the windows could be opened. The bodies were rectangular with a curved roof profile. The cars seated 30, all third class. The seats were reversible, tramway style.

The livery was GSR red, with the company's crest on the centre panel. The fleet number was carried on the standard gauge cars just below the window on the last panel before the left-hand door, while on the narrow gauge cars it was carried to the left of the passenger door below the grab handle.

The inspection cars could seat about eight people and, like their larger cousins, could be driven from either end but they were never used for passenger carriage. They had various engines over the years, but always petrol. Externally, they resembled railcars 385-6 and 395-6, but were shorter with only three window bays (see page 114). Their livery was teak with gold, black and red-shaded lettering.

After delivery of the two narrow gauge cars in late 1927, three drivers were trained specifically to drive them. One car entered service during the winter and spring periods, when loadings were light, on the Kilkee to Moyasta branch. It also made one through-run daily to Kilrush with school children. The other car worked the down morning and up mid-day service all the year round. They were able to haul a small four-wheeled coach which was used to carry luggage, and the guard! They were not able to cope with heavy traffic, and at such times steam trains were used.

Great Southern Railway narrow gauge Drewry petrol railcar No 396.
National Railway Museum

Great Southern Railway broad gauge Drewry petrol railcar No 385.

National Railway Museum

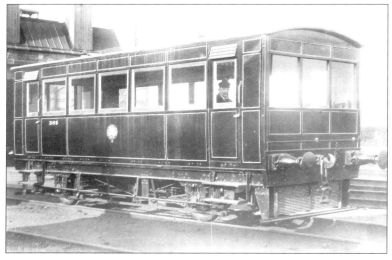

Working the main line proved to be unsuccessful and only lasted two seasons. The cars broke down frequently, even on the branch line where they were more lightly worked. They continued to be used on the branch until 1936 after which they were withdrawn and lay out of use at Ennis. They were moved to Inchicore in 1939 and were withdrawn and scrapped there in 1943.

The two standard gauge cars were introduced into service in 1927-28. As with the earlier steam railmotors and railcars they were used on the Goulds Cross-Cashel branch. Later, No 386 was put into service as a track inspection vehicle.

Subsequently, in 1930 No 386 was modified in Inchicore as a test bed for the Drumm battery railcar experiment. It had its bodywork modified at one end into a streamlined form and was equipped with an experimental 110 volt Drumm battery and two 30 hp electric motors.

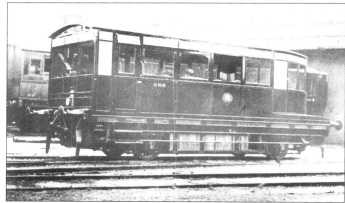

GSR Drewry railcar No 386 converted to a test bed for the Drumm battery project. National Railway Museum

In a series of tests conducted between Inchicore and Hazlehatch, a distance of about eight miles, it was able to achieve speeds of 50 mph and, after exhaustive trials, it was decided to build a full-sized set of battery driven railcars. This development is covered more fully later in this section (page 95).

After these experiments were completed, the modified railcar was put into service between Inchicore and Kingsbridge (Heuston) as 'The Cab',

transporting railway staff between the city and the works. Both cars were withdrawn in the 1940s.

The inspection cars had much longer lives. They survived into CIÉ ownership and were replaced by new inspection cars in 1964. No 2 remained in teak livery until scrapped, but Nos 3-5 got CIÉ green in the 1950s. One, possibly No 4 in Limerick, became black and tan after 1962.

Diesel railcars, 1931-39

CDRJC

In 1930 the CDR decided to power all future railcars with Gardner 6L2 diesel engines and Nos 7 and 8 appeared in this form. They were notable as being the first diesel railcars used for regular passenger services anywhere in Ireland or Great Britain.

Again they were a GNR design, with O'Doherty of Strabane bodies, and were 27'0" long and 7'0" wide and carried 30 passengers. They had a maximum speed of 43 mph and, on a closely monitored trial run, No 7 returned a fuel consumption of 25 mpg.

The engine was mounted vertically, in line, forward of the cab and drive was taken to the two rear axles which were connected by a chain drive. The front axle was supported on a pony truck which protruded in front of the radiator, resulting in a rather ugly front end to the vehicles. However, they were very successful and ran for 18 years, and confirmed the use of diesel engines on all the company's future railcars.

Railcar No 12 was the next big innovation after the introduction of diesel power in Nos 7 and 8. No's 9 and 10 in the railcar sequence were railbuses (see page 107) and No 11 was the locomotive *Phoenix*, a conversion from a CVR steam tractor. This had been bought from the CVR in 1932 and fitted with a Gardner 6L2 diesel engine. It gave many years useful service, mainly on shunting duties.

The design of No 12 and all subsequent CDR railcars was influenced by the design of CVR railcar No 1 (see page 86). The power bogie was built by Walker Bros of Wigan, with bodywork by the GNR in Dundalk. It was different from previous railcars, in that it had an articulated chassis, of a type which Henry Forbes had first seen on CVR railcar No 1, a car with which Forbes had been very impressed. On No 12, the 6L2 Gardner 74 hp engine and driving cab were mounted on the power bogie. This was articulated to the 41 seat coach section which was carried on an unpowered bogie. Roller bearing axle boxes were fitted and the driving wheels had coupling rods like a steam locomotive. The body

CDR railcar 7 or 8 at Donegal Town.   Railfotos

CDR railcar No 12, with specially built railcar trailer, at Strabane. Railfotos

was 31'0" long by 7'0" wide. No 12 has survived into preservation as part of the Foyle Valley Railway collection.

Railcars 14 and 15, delivered in 1935 and 1936 respectively, were similar to No 12, but No 15 had a full width cab, unlike the half-cabs of 12 and 14. Nos 16, 17 and 18, supplied in 1936, 1938 and 1940 respectively, were basically the same as No 15, but were powered by 102 hp Gardner 6LW engines and were slightly heavier. They were very successful, 16 and 18 lasting until the closure of the rail network, No 17 having been destroyed in a collision in 1949. No 18 was damaged in 1949 by fire, but was rebuilt in Dundalk with a new body, similar to Nos 19 and 20. No 18 is preserved by the Foyle Valley Railway in Derry. It was rebuilt again in 1997-8 while in FVR ownership.

The CDR railcars were capable of hauling trailers to extend their passenger carrying capabilities. Henry Forbes always had an eye for a bargain and given the relatively impecunious state of the CDR, this was a useful talent in its manager. (Whether the CDR should have had to operate within such a strict financial straitjacket, given that it was jointly owned

by the relatively prosperous LMS (NCC) and GNR is another question!) As the early railcars were withdrawn, some were rebuilt as trailers. Some vehicles were bought or built as lightweight railcar trailers and, as we have seen, others were converted from railcars formerly operated by other railways.

All CDR railcars and passenger trailers were finished in a very attractive geranium red and cream livery and carried the CDR coat of arms on the side.

CVR railcar No 1 and diesel tractor No 2

The CVR was essentially a road-side tramway, and was so called for the first seven years of its existence until 1894, when it changed its name (but little else) from 'Tramway' to 'Railway'. By 1928 its slow steam tramway operation was coming under severe pressure from road competition.

In that year, the management of the Clogher Valley Railway was taken over by a Joint Committee appointed by the county councils of Fermanagh and Tyrone. The reason for this was an attempt to improve the finances of the company which had degenerated from poor, which they always had been, to critical. The new management team included Henry Forbes,

CDR railcars No 15 and No 10 at Donegal Town. No 10 was originally Clogher Valley Railway No 1.   Railfotos

the general manager of the CDR.

In an attempt to improve timings and reduce operating costs, the CVR purchased a diesel railcar and a diesel goods tractor from Walker Brothers of Wigan. This was to allow a cut back on steam operations. This order was placed following the successful trials of CDRJC railcar No 4, which had been loaned to the CVR.

Railcar No 1, which went into service in 1932, was the first articulated, power-bogie railcar to run in Ireland. Its 28 seater body was built in the GNR shops at Dundalk and was mounted on an unpowered bogie. The engine was a 74 hp Gardner 6L2 diesel. It had a half-cab and engine assembly mounted on the four-wheeled power bogie. The front driving wheels were connected by coupling rods.

No 2, the diesel goods tractor, which had been delivered in 1933, comprised a Walker half-cab body with a Gardner 6L2 diesel engine, but with a six-plank, drop sided flatbed lorry assembly mounted behind the cab. The whole vehicle was carried on a four-coupled power bogie unit similar to that fitted to the railcar. The driver could enter the cab only through a door located in the centre of the back

bulkhead cab , just as in the railcar. This meant he had to climb over the flatbed body and any contents it might happen to be carrying! This tractor was capable of hauling carriages or vans.

Like the railcar, it too was bought by the CDRJC on the closure of the CVR. It does not seem to have run on the CDR, but was cannibalised to provide parts for the other Walker railcars. In fact it is thought that the vehicle now in the Ulster Folk and Transport Museum (UFMT) comprises the railcar passenger body and the tractor power bogie and cab.

Although the diesels achieved their aims, road transport competition proved to be too strong and the railway was closed in 1942.

The pioneering CVR railcar No 1, which Henry Forbes had admired in 1932 and which influenced subsequent CDR railcar design, came to the CDR in 1942, when the CVR closed. It was given the number 10. The power bogie was similar to that on No 12, but the body was shorter and could seat only 28 passengers.

It lasted until the end of CDR rail operations and is now preserved in the Ulster Folk and Transport Museum at Cultra, Co Down.

The GNR railcars

As has been outlined above, the CDRJC introduced light railcars to fight back against road competition. The CDR was jointly owned by the LMS (NCC) and the GNR. It was administered by a committee appointed jointly by these two concerns, the Joint Committee in the CDR's title. As a result, railcar design and construction was carried out in the GNR's Dundalk workshops.

With the experience so obtained, the GNR decided to construct similar vehicles for use on its own lines. This was for much the same reasons as the CDR, to compete with growing road competition. This was developing into a serious problem in the GNR operating area, where the roads were generally better than in the CDR's hinterland in the rugged north west of Ulster.

The GNR's highly successful vehicle development programme comprised three main elements: firstly, the construction of railcars for branch line working, the subject of this section; secondly, the conversion of some road buses to railbuses for very lightly trafficked branches; and thirdly, the development of multi-unit diesel trains for main-line working. As a result of these developments, the GNR was a pioneer in the British Isles in the use of diesels, particularly for light branch line work.

GNR railcar A leaving Banbridge.   RC Ludgate, author's collection

GNR railcar B at Scarva.   Real Photographs X 124

Experience with building the CDR's railcars allowed the GNR cars to be built strong but light. Frame and body weights were kept well below that of conventional steam-hauled stock. This resulted in excellent operating costs. All railcars were finished in the new and attractive GNR Oxford blue and cream livery, rather than GNR coach mahogany.

The early railcars were not always well received by steam locomotive men and there are stories of the railcars being sabotaged by having engine oil sump plugs removed during the night, or fitters having cinders thrown at them from passing steam locomotives, or water hydrants in engine sheds mysteriously 'turning themselves on' and drenching fitters working under the railcars.

In 1932 railcars A and B were introduced. Their specifications had been drawn up in consultation with the traffic department. A 50 mph maximum speed was required for a car operating on the level, without a trailer. If hauling seven tons, the weight of a loaded wagon or horsebox, speed was to be 40 mph, or hauling two such loads, 35 mph. The cars had to be capable of being driven from either end and had to be fitted with regular buffers and draw gear. Each car was to seat 32 passengers. Great care was taken in the design of the body and frames to keep overall weight down.

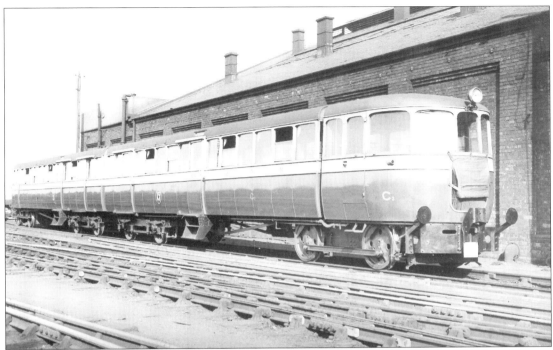

GNR twin railcars C2 and C3 at Dundalk.   National Railway Museum

Railcar A was fitted with a 130 hp AEC power unit, had a diesel-mechanical drive and weighed 18¾ tons. Railcar B had a diesel-electric drive comprising a 120 hp Gleniffer engine linked to a Tilling-Stevens generator. The whole vehicle weighed 21 tons. Railcar A was later fitted with a 102 hp Gardner 6LW engine and railcar B with a 6L3 Gardner engine.

Both cars were put to work in the Portadown area and were monitored closely. They gave fuel consumption figures of 8-10 mpg and operating costs of under 4d per mile – very favourable figures indeed. Railcars A and B worked mainly on the Scarva to Banbridge branch. Under UTA ownership, A also worked on the Derry-Strabane and Derry-Omagh runs. In 1946 railcar B was converted to a locomotive hauled passenger trailer and renumbered 500. It was taken out of service in 1949 and lay at Dundalk until it was eventually scrapped in the mid 1950s.

Railcars C, C₂ and C₃

In 1934 railcar C was introduced, based on the lessons learned from the tests of A and B. Railcar C was a very different design from that of its two predecessors. It had its driving compartment and engine separate from the passenger section and articulated to it just like the CDR railcars. Buffers and draw gear were provided at both ends of the vehicle.

The driving and passenger compartments were linked through a connecting door. Passenger capacity was 50 in one class of accommodation. The whole design was governed by the need to get weight down to the absolute minimum.

The engine fitted was a 96 hp Gardner water-cooled diesel, operating a mechanical drive system via a four-speed gearbox. It returned a fuel consumption of about 12 mpg. As with the CDR railcars, the front four driving wheels of railcar C were coupled. The coach section was carried on a plain four-wheeled bogie. The railcar was fitted with a vacuum brake system operated from an exhauster driven from the engine. The whole vehicle weighed 14 tons 5 cwt.

Unlike A or B, railcar C could only be driven from one end and had to be turned on a turntable at the end of a run. It was used on the Enniskillen-Bundoran branch and was required to run nearly 1,000 miles in a six-day week. It was also used later on the Cavan to Clones branch and for the mail run between Dundalk and Cavan, after passenger services had been withdrawn in 1957.

To get round the problem of being single-ended, the next two railcars, C2 and C3, introduced in 1935, were coupled together back-to-back. Unlike C, which had a curved rear section, C1 and C2 had flat rear ends to allow back-to-back coupling, though there was no corridor connection between them. Otherwise, they were similar to C (now renumbered C1). At 15 tons, they were also slightly heavier than C1. They could be used separately if required but, when used together, only one car was under power, the other being towed in neutral gear.

The engines fitted to these cars were Gardner 6LWs of 102 hp. Maximum speed was 46 mph. The seating capacity of C2 was 52 (later 48) and C3 seated 46 in three classes and there was space for about ¼ ton of luggage. Both bogies and layout were by Walker Brothers of Wigan. As a double unit, they were used on the Dublin-Howth and Dublin-Balbriggan services. After 1937 they were run separately, mainly on the Irish North services.

As a result of experience with these early railcars, the Great Northern's engineers came to the conclusion that units of this type must be designed for a specific service and to use them for general or varied traffic was out of the question. The first requirements of any design were to be dictated by the traffic department – passenger capacity, speed, etc. Once these were known the ideal railcar would combine these with a minimum possible weight in relation to the necessary power. This logic then influenced subsequent GNR railcar development.

Railcars D, E, F and G

It was found from experience that neither the coupling of two single units, A and B, nor the joining of the C-type articulated units gave the best results, so the company settled on a new standard of design which led to the production of Railcars D and E in 1936.

Railcar D was the first twin coach,

centre-power-bogie, railcar. The engine was operated by remote control from cabs situated at each end of the two passenger cars. The centre section contained the engine, a Gardner 6L3 153 hp diesel with a fluid-flywheel attached, coupled to four-speed epicyclic self-changing gearbox which was in turn connected via a 'reverse-box' to one of the three axles under the centre section. Top speed was 42 mph.

The centre bogie had a wheel arrangement with three coupled driving wheels on each side. Because of the length of this bogie, the centre wheels were flangeless. The centre compartment was divided into two parts. One part contained the engine assembly, the other the guards van, under the floor of which was the reverse-box lever. This lever was spring-loaded and pin-locked and had three positions; reverse, neutral and forward. At each terminus, the driver moved the lever to the extreme opposite position for the return trip. Gearbox and transmission shafts were located under the floor.

At each end of the centre bogie, the inner ends of the coach sections were pivoted on ball-and-socket joints. Centre section and coach bodies were joined by standard corridor mountings. The outer ends of the coach bodies were mounted on plain bogies.

The driver's cabs were situated in the centre of the front of each coach, with passenger seats on either side. Controls consisted of a hand throttle, dead-man's handle, vacuum brake, hand brake and manual windscreen wiper lever. One big change for drivers, compared to the earlier railcars, was the fact that gear changing was automatic.

The passenger capacity was 159 – eight first-

GNR railcar E interior. Duffner brothers

class, 50 second class and 101 third class. The weight was 39½ tons. The layout was again by Walker Bros, and gearboxes by the Wilson Self-changing Gear Co; the GNR workshops, Dundalk, built the bodywork.

Railcar D proved to be very economical and had a high availability. Railcar E, a similar vehicle, quickly followed it into service a month later. They were used initially on the Dublin-Howth service.

Later they went north, one working the Newry-Warrenpoint branch and the other in the Belfast area.

In 1938 broadly similar railcars, F and G, were introduced. As with D and E, the new cars were articulated but now with a twin-engined, twin-axle centre power bogie. Each engine drove through its own fluid-flywheel, self-changing gearbox and reverse box on to its own axle, the one furthermost

GNR railcar E leaving Amiens Street station, Dublin, in 1948   HC Casserley

GNR railcar G at Queen's Quay station, Belfast, on railmotor replacement duties.

RC Ludgate, author's collection

from the engine. Each engine was equipped independently with a full set of ancillary equipment and there were, in fact, two power units on one bogie. The engines fitted in opposite corners of the centre section, at right-angles to the axles. They were synchronised at their throttles and could be driven together from either cab.

Transmissions were below floor level but it still was possible for the guard to walk through the engine compartment to the coaches. Coaches and cabs were the same as on D and E, except that the guard's van was now in one of the coaches rather than in the centre section, as had been the case in D and E. The railcars each seated 164, weighed 41¼ tons and had an overall length of 125' 9½".

They had more rapid acceleration and higher top speed than D and E and excellent pulling power. As a result, they were used not only on Dublin-Howth and Dublin-Drogheda work, but even on Dublin-Belfast specials. The engines were 102 hp Gardners, with layout by Walker Brothers, gearboxes by Wilson Self-changing Gear Co and bodywork by the GNR, Dundalk works.

During World War II, when coal supplies were more restricted than oil, this railcar fleet allowed the GNR to provide a higher level of service than otherwise would have been possible. Their work was doubled and loads of 400 passengers were recorded on railcars F and G.

The railcars also helped retain many branch line services long after they might otherwise have closed in the face of road competition. They were also the precursors of the post-war multiple-unit diesel trains on the GNR.

After the Ballymacarrett accident on the BCDR, GNR double-ended railcar G was used to replace the push-pull steam train which had been involved in the accident.

On the break-up of the GNRB in 1958, railcars A, C3, D and F were allocated to the UTA, which used them on various parts of its of its shrinking system. The other cars – C1, C2, E and G went to CIÉ.

Railcars A, C3, D and F were given the UTA numbers 101-104 respectively. Nos 101 and 104 were repainted in the UTA's dark green livery, with yellow and black striped front warning panels. Railcar E was later sold to the UTA by CIÉ in 1960. It was not given

a UTA number and was cannibalised to keep No 103 running. Railcar G was also sold to the UTA in 1963 and became UTA No 105.

Railcar A was earmarked for preservation by the Belfast Transport Museum, but was withdrawn after being damaged in a shunting accident with locomotive No 33 at Foyle Road, Derry, in October 1962. It was sold to the contractor who lifted the GNR 'Derry Road'. He had the body removed from its bogies and mounted on a carriage underframe as accommodation for the workers. Of the original vehicle, only one cab remained on the underframe and was used as a tractor unit.

On completion of this contract, railcar A lay in Portadown, out of use until the same contractor used it again in 1967 to lift the Mallow-Dungarvan line in the Republic of Ireland. The remains of railcar A were broken up in Dungarvan at the termination of this contract. Railcar F was also involved in the lifting of the 'Derry Road' and was subsequently sold to the contractor lifting the Clonmel to Thurles line. It was still in use as late as 1970. Cars C1, C2 and C3 were withdrawn in 1961, D in 1963 and E in 1960.

The LMS (NCC) railcars.

In 1933, close on the heels of the withdrawal of the Sentinel steam railcar, the LMS (NCC) produced their railcar No 1. This twin bogie vehicle was constructed at York Road, the NCC workshops in Belfast, and was intended for use on branch lines. The car was designed to be capable of hauling a light load. It was the first of two railcars ordered for this purpose by Mr Speir, the NCC manager.

Railcar No 1 was really a standard coach powered by two vertical, 10-litre, 125 hp, petrol engines mounted under the floor at the centre of the railcar. The petrol engines were replaced in 1947 by two 130 hp Leyland diesel engines. It was again re-engined with two Leyland 0600 125 hp diesels in 1959. Power was transmitted to the wheels through a Lysholm-Smith torque-converter, direct drive being engaged at a track speed of about 25 mph. As electro-pneumatic controls had not, at this stage, been developed, throttles, clutches, and final drives were operated by hand. The vehicle was fitted with vacuum brakes, vacuum being created by a motor-driven exhauster and maintained by an auxiliary exhauster driven from the transmission.

LMS (NCC) railcar No 1. It was originally fitted with two petrol engines, which were later replaced with diesels.

Author's collection

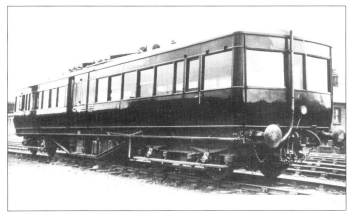

The length of the railcar was 57'0", width 9'8" and weight 32 tons. It could be driven from either end and had seats for six first and 55 third class passengers.

During the next few years, three more railcars were introduced. The three new cars all had raised driving positions to allow them to propel trailers, unlike No 1 which had to run around its trailer at a terminus.

Railcar No 2 appeared in 1934. Unlike No 1, it was diesel powered from the outset, being fitted with two 130 hp Leyland engines. Like No 1, it was a double-ended, twin bogie vehicle and had a lightweight integral aluminium body, 60'0" long. This gave it a weight of 26 tons, six tons lighter than No 1.

The body design of No 2 was not very attractive, being very angular and the raised driving positions were of a particularly ugly design. These were later removed and a more normal cab arrangement was substituted. No 2 had a similar motive power and transmission configuration to No 1, but had hydraulically operated throttles and an electro-vacuum system for operating the clutches and reversing mechanism. It was fitted with a vacuum brake system similar to that fitted to No 1.

Although designed for branch line working, both 1 and 2 were so successful that in fact they spent most of their time working stopping trains on the main line and the Larne line.

Railcar No 3 appeared in 1935 and No 4 in 1938.

LMS (NCC) diesel railcar No 2, as originally built. The elevated driving positions were later removed.

National Railway Museum

LMS (NCC) railcar trailer No 1 in LMS livery in 1947. The vehicle seems to be still fitted with some wartime blackout windows.

RC Ludgate, author's collection

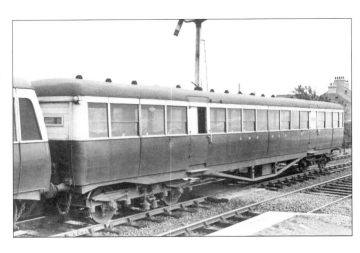

These were more sophisticated than their two predecessors. The design of Nos 3 and 4 was much more attractive than that of No 2, with a more streamlined body design and the raised driving position properly integrated into the overall body shape (see page 117).

The technical details for Nos 3 and 4 were the same. Length was 63'6" over buffers with a 62'0" body and width was 9'6". Height was 12'5½". Two Leyland 130 hp diesel engines were located under the floor in the middle of the car. Transmission was through a gearbox and torque-converter to the inner axle of each bogie truck. Weight was 28.7 tons. Vacuum brakes were fitted. They each had seating for 72 passengers – 12 first class and 60 third class.

Two special lightweight trailers were built in 1934 to run with railcar Nos 2-4, which considerably enhanced their passenger carrying capabilities. For example, No 3's own capacity of 72 was increased to 272 when used with two lightweight trailers, one fore and one aft. The 100 seater railcar trailers were 62'0" long by 9'5" wide and weighed 17½ tons. In the UTA renumbering of 1959 they became Nos 544 and 545.

Propelling a trailer did restrict the driver's vision and was eventually prohibited after a railcar propelling a trailer hit a cow! On other trains, the railcars were used to pull vans or horse boxes.

Of the four railcars, No 1 survived in service until 1965, when it was set aside for preservation; No 2 was withdrawn in 1957; and Nos 3 and 4 were destroyed in fires, the former in 1957 and the latter in 1968, although it had been out of service since 1965.

All four railcars were originally finished in LMS crimson lake livery, although they were repainted into red and cream in the late 1940s. When the UTA acquired them, on taking over the NCC in 1949, they were repainted in its dark green and cream corporate livery. The UTA green went through various permutations and No's 1 and 4 eventually had all-over green with a striped warning panel.

The Drumm battery trains

In 1929 a group of scientists at University College Dublin, headed by a Dr JJ Drumm, patented an improved type of alkaline battery. The main advantage of this battery was that even at high rates of discharge, the voltage of each cell was 50% higher than that of the standard Edison cell. As a result, the Drumm battery could be charged and discharged, without damage, many times each day, at rates much greater than those that were normal at the time. The development of this battery was an extremely innovative engineering project. It was claimed that the battery could stand comparison with the capabilities of the petrol engine.

Drumm approached the Irish government for help in assessing the battery's potential with a view to ultimate commercial development. The government saw the chance to use the electricity generated from the new Shannon hydro-electric scheme as a possible way to rejuvenate the newly amalgamated Irish railway network. It granted funds to allow the claims for the battery to be assessed, with future funding being dependent on the results.

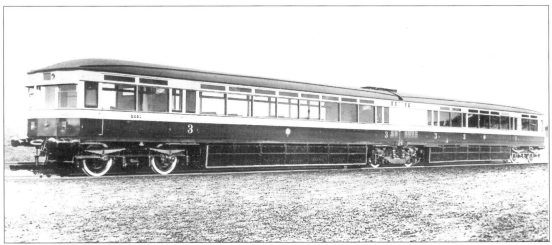

Drumm battery articulated railcar A, carrying its original number 2501.   National Railway Museum

The GSR made development facilities available at Inchicore.

As noted above, Drewry four-wheel petrol railcar No 386 was converted as a test bed for the new battery (see page 84). Two 30 hp 110 volt DC axle-driven electric motors with associated switch and control gear were fitted. The 110 volt Drumm battery was constructed in Inchicore works and was fitted between the wheels of No 386.

On 21 August 1929 trials began on the Great Southern main line between Kingsbridge and Hazlehatch, eight miles away. It was clear from these trials that this was an exceptional new development in battery-electric locomotion. Speeds of 50 mph and acceleration figures of ¾ mph/sec were maintained without any strain on the battery.

As result of the success of these tests, it was decided to build a full-sized two-car articulated unit to seat 140 passengers. This unit, designated 'A', was built in Inchicore and went into service early in 1932, running between Amiens Street (Connolly) and Bray, with occasional trips to Greystones. Railcar A ran about 134 miles per day, making stops at 14 intermediate stations. It achieved an acceleration of 0.85 mph per sec and a maximum speed of 47 mph. Charging stations were installed at Amiens Street and Bray. These allowed the batteries to be given a boosting charge at the end of each trip.

A second two-car articulated unit, railcar B, with an improved battery, was introduced later in 1932 on the Bray service and railcar A was taken out of service to have the improved batteries fitted.

In the summer of 1932 a special train was run for President de Valera and members of the government. It was to cover the line from Dublin to Gorey in Co Wexford and back, a round trip of 120 miles over difficult track. The outward trip was made without incident but, on the homeward trip, the planned 30 minute boost at Bray had to be cut short to 10 minutes because of de Valera's impatience to return to Dublin. As a result, the train only made it back to Westland Row station rather than Connolly. However, this trip was regarded a success and the trains went into regular service.

Each train ran seven days per week, covering about 1,000 miles per week. In order to handle larger numbers of passengers, the two sets A and B were modified to run in multiple. In addition, a specially wired bogie coach was constructed to be intersposed between the two trains to allow the formation of a five car train. This could seat over 400 passengers.

By 1935, the GSR had spent just over £27,000 on the Drumm trains. This can be compared with a cost of £20,000 incurred in 1936 for five new 4-4-0 steam engines. However, by that year the two Drumm trains had covered over 400,000 miles between them.

Two further railcars, C and D, were built at

Inchicore and went into service in 1939 on the Harcourt Street-Bray route. These had a more streamlined design, with better designed seating. This allowed 13 weekday departures of Drumm trains from Harcourt Street and a consequent reduction in the number of steam trains required.

The trains belonged to the Drumm company and a rental was paid by the GSR for their use. An attempt was made to interest the GNR in the trains. A trial was arranged between Amiens Street and Howth, but although the GNR people were impressed, nothing further came of the demonstration. Cars C and D were later tried on the Cork-Cobh and Mallow-Tralee branches.

During the 1939-45 'Emergency' the four trains provided a much needed service, at a time when private transport was almost at a standstill and only a skeleton steam service was possible because of fuel shortages.

A proposal, which never progressed beyond the early planning stage, was made to build an express locomotive on the Drumm principle, for use on the Dublin-Cork line. It was planned that a 30 minute boost charge would be given at the half-way mark. To avoid this 30 minute stop, an alternative proposal was to electrify a 20-mile stretch of track, thus allowing the battery boost to be taken on the move.

In 1942, a paper was read to the Institution of Electrical Engineers on the subject of electrification of the railways in Ireland. The thesis put forward in this paper was that although electrification of the Irish railway network was desirable, it was not economically feasible using either overhead or third-rail systems, but that electric locomotives using the Drumm system would be feasible. Short stretches of conductor third rail would be installed to assist acceleration. For branch line trains, Drumm battery railcars would be used.

Dr Drumm, working in America, had further developed his battery. The ratio of weight to output had been reduced by 20% and the bulk by 10%. This would have allowed trains to operate faster and further. However, a proposal to allow further government funding to develop the electrification of the Irish railway network on the Drumm system, was turned down because of the 'Emergency'.

By 1950 the batteries were worn out and the cars needed a major overhaul. However, the decision was taken by the newly formed CIÉ not to proceed with this. As a result, the railcars were withdrawn and converted into steam-hauled coaches. In this form they had a short life, being replaced by new AEC diesel-mechanical railcars running on cheap post-war oil. Shortly afterwards, one of the Drumm train routes – the Harcourt Street-Bray line – was closed.

Passengers on the Amiens Street-Bray line had to wait for the introduction of the DART trains, before the quality of Dr Drumm's railcars could be experienced again. If it had not been for the war and a shortage of money for further research, the Drumm battery trains might have allowed Ireland to become a world leader in the development of battery-powered rail traction.

The equipment fitted to the experimental Drewry car was a 110 volt Drumm battery, two 30 hp traction motors built by the Victory Dynamo Co of Leeds and control gear by Valsto, Clarke and Watson.

Railcar A was powered by two 300 hp 600 volt DC motors driving the axles of the central bogie. It weighed 85 tons, including 13.5 tons of battery. It was 126 feet long and could seat 140 passengers in open saloons. It was the most powerful battery-driven railcar constructed up to that time and was capable of keeping the same schedules in suburban rail services as a similar vehicle operating on a normal electrified system. Later cars were of basically the same design

Charging stations, running off the national grid, consisted of a transformer and a mercury vapour rectifier. Leads from the charging station were brought out over the track and pantographs, raised and lowered with air pressure, made contact to allow the charge to be taken. The battery instrument panel was located in a the guard's compartment.

The livery of railcars A and B was originally GSR lined chocolate, with cream upper panels. The GSR crest was carried half-way along the sides of each coach, with the class designation in large numerals on the doors. The railcar letter was carried centrally on the front and rear over the driver's window. Later, the livery was all-over GSR lined maroon. C and D were introduced in this livery This was later replaced on all four railcars by the CIÉ two-tone green 'flying snail' livery.

Post-war railcars

SLNCR railcar B

The Sligo Leitrim and Northern Counties Railway was a small standard gauge line linking Enniskillen, on the Great Northern system, with Sligo on the Midland Great Western (later part of the Great Southern Railways and finally of CIÉ). It traversed fairly sparsely populated country and the bulk of its traffic was cattle, passenger traffic representing a relatively small proportion of its custom. Like the GNR, the CDR, the LLSR, and the DNGR, the partition of Ireland in 1921 adversely affected the operation of the railway. The Sligo Leitrim also faced increasing competition from road transport.

In 1932 the SLNCR tried out a GNR railcar on its line. It is not clear now whether this was railcar A or B, both of which came into service with the GNR in 1932. However, the railcar's attractive operating characteristics encouraged the SLNCR to order their first railbus, in 1934, from the Great Northern. Railbus A was delivered in 1935 and was followed over the next few years by others, as described in the chapter on railbuses.

In 1947 the SLNCR took delivery of a railcar from Walker Brothers of Wigan (photo page 117). This railcar, which was designated B, was purchased as the result of the company's desire to reduce the operating costs of its scant passenger traffic. The operating economics of the railbuses had persuaded the directors of the permanently financially strained SLNCR, to invest £10,500 in a larger purpose-built railcar, similar to the ones which had been such a success on the CDR and GNR

Railcar B was powered by a 107 hp Gardner diesel engine, mounted on a four-wheeled power bogie. On this was constructed the forward driving cab which enclosed the engine assembly. The power bogie was of the four-coupled wheel arrangement with outside rods. It was articulated to the main passenger coach, which was carried on a plain bogie. Transmission consisted of a fluid flywheel, a Wilson epicyclic gearbox, propeller shaft and an air-operated final drive and reverser unit. The railcar was 54'1¼" long, 9'6" wide and weighed 18 tons 12 cwt.

Maximum speed was 45 mph. It returned a fuel consumption of 12 mpg and operating costs of 4d per mile, one eighth of those of a steam train.

It could accommodate 59 passengers in a two plus three seating arrangement. Unlike the Donegal and the GNR C-class railcars, also Walker Brothers designs, railcar B could be driven from either end. There was a full cab at the engine end and a half-cab was set into the rear of the coach section.. The vehicle was, for its time, modern, comfortable, attractive looking and was well liked by both passengers and staff.

When the SLNCR closed in 1957, railcar B was bought by CIÉ and became railcar 2509. It was used for driver training, light passenger work and a few enthusiasts' railtours. It was finally withdrawn from regular passenger workings in 1970-71, its last duties having been on the Limerick-Nenagh line, and ran its last railtour, for the Irish Railway Record Society, in 1971. After having been stored at Mallow for several years, it has recently been recovered by CIÉ with the intention of making it an exhibit at the new working railway museum to be established at Mullingar.

Railcar B's SLNCR colour scheme was the two-tone green scheme also applied to the company's road and railbuses (page 114). When in service with CIÉ, it carried that operator's green livery up until 1962 and after that date the black and orange livery.

CDR Nos 19 and 20

The last railcars built for the CDR were No 19, delivered in 1950, and No 20, delivered in 1951. With the cab units built by Walkers, fully enclosing the engine, they looked much more modern than the previous vehicles. Again, the 41-seater body was by the GNR in Dundalk. (Four similar railcars were built for the West Clare Railway.) However, they could only be driven from one end.

When the CDR rail system was closed, these County Donegal railcars were bought in 1961 by the Isle of Man Railway Co and they still survive there. They have recently been fully rebuilt at the company's workshops in Douglas.

The UTA's single-unit railcars

The UTA introduced four self-contained railcars. The first was the Ganz railcar (No 5) acquired from England in 1951. The other three were built by the

CDR railcar No 19 and trailer at Strabane. This car is now in the Isle of Man.   Author's collection

Authority in 1961-2, as part of its Multi-Purpose Diesel train fleet.

The Ganz railcar had its origins before World War II. The Hungarian State Railways introduced the 'Arpad' diesel railcar, built by Ganz of Hungary, for use on a 56 mph service between Budapest and Vienna. The Metropolitan Vickers Electrical Co of Birmingham, together with the Metropolitan Cammell Carriage and Wagon Co, obtained a license to build the Ganz design in England, with a Metro Cammell body. This was in response to a rapidly increasing demand from South America for this type of vehicle. The Ganz license was seen as a way of gaining increased access to this market.

A prototype car was erected in Birmingham and, after several runs over LMS lines, the railcar, now known as the 'Metro-Vick Cammell Car', was publicly unveiled on 30 June 1937. It was tested on the LMS line from Euston to Tring, completing the 31¼ mile trip non-stop in 30 minutes, with long stretches at 70 to 74 mph.

After this test run and publicity, the Ganz car dropped out of public view, although it does seem also to have been tested in the Hull area before the war. It was probably put into store in Birmingham, by

Metropolitan-Cammell, during the war. There is some evidence that it ran again on tests in the north-east after the war.

The Ganz car was the first British railcar to have a large bogie-mounted engine and the first to have a mechanical transmission with an engine over 200 bhp. The 240 bhp Ganz engine was mounted on one bogie and projected high above the floor of the railcar. Its transmission was built into the leading bogie, both axles of which were driven.

The engine compartment was encased in polished wood, looking a bit like a baggage compartment, around which there was a passage way. The casing was double-walled with an airspace in between, the exhaust passing up to the roof through a 'Burgess' silencer. Fuel and water tanks were arranged separately inside the compartment, strapped on the wall/ceiling above the engine casing. The interior panels were of mahogany plywood "polished in the natural shade". All window and door frames were bent walnut "polished in a rich brown shade". The floor was of tongue and groove board and was covered with cork carpet and rubber. The car was 64'0" long by 8'7" wide and weighed 38 tons.

In February 1951, as part of its dieselisation

UTA 'Ganz' railcar No 5 at York Road in 1957. Warning stripes were later added.   N Simmons

programme, the Ganz railcar was purchased by the UTA. Before delivery to the UTA , new 5'3" bogies were fitted in England. The Ganz was the first vehicle using air brakes to run on the UTA . As UTA No 5 it was used on the Coleraine-Portrush shuttle service and on other parts of the UTA system, including the ex-GNR 'Derry Road'.

No 5's seating capacity was 18 first class and 38 third class, but ran with a specially built trailer, No 215, to increase capacity. This trailer was built in 1953 specifically for use with the Ganz railcar. In 1959 it was renumbered 515. After withdrawal of the Ganz, the trailer was modified in 1968 and transferred to the Multi-Engined Diesel railcar fleet.

The UTA colour scheme for No 5 was Brunswick green and cream. This unique railcar was withdrawn from service in 1965 and scrapped the same year at Maysfield yard in Belfast.

In 1952-53 the UTA built a fleet of multi-engined diesel railcars (MEDs – really DMUs) for use on its suburban lines and in 1956 decided to expand the use of diesels on the ex-NCC section, by introducing two high-speed trains for use on the Derry line, each to have four single-engined power-cars and a non-powered dining car.

The UTA wanted units which would be capable of

handling these Belfast-Derry expresses, running at speeds of up to 75 mph in the daytime, and freight trains of up to 100 tons per power unit, at night. They were to perform light shunting duties as well!

Thus, maximum flexibility of operation was the overriding factor in drawing up the new design. For this reason they were named Multi-Purpose Diesel trains, or MPDs. The bodies on MPD units were based on conversions from existing carriages and, in building the MPDs, as much as possible of the original coaches was retained. In 1957-59, 27 MPD power cars (Nos 36-62) were put into service.

Each car had one six-cylinder engine, a Leyland 0/900, turbo-charged to 275 hp. The hydraulic transmission was by Self-Changing Gears Ltd, with a Schneider torque-converter attached. Gears were totally automatic, the driver controlling forward and reverse throttle and brakes only. Although each car had only one engine, both axles of the trailing bogie were driven. MPD stock were fitted with air brakes, but were supplemented by equipment to deal with the haulage of vacuum-fitted stock.

The Leyland engines originally fitted were never totally satisfactory and, in 1964-69, this fleet was re-engined with Rolls Royce and AEC units.

One car, No 58, was written off in a level-crossing

UTA single unit MPD railcar No 63 approaching Whitehead.   Official UTA photograph

accident near Downhill in 1959, after only six weeks in service. Resulting from this, it was decided to build, not only a replacement, but two additional units. These three cars were constructed in 1961-62 as Nos 63-65. Unlike the earlier cars, they were double-ended, with left hand drive, rather than right hand. They could therefore operate as single units and were useful for shunting work and light passenger turns. They were used, amongst other things, on the Coleraine-Portrush shuttle. On weekdays a car would operate as a single unit, but on Saturdays a trailer was added to give extra capacity.

Withdrawals of MPDs began in 1974. They had been moved off the Derry services, and concentrated on the Larne line and secondary services, when the diesel-electric 70-class units appeared in 1966-67. They were introduced briefly and unsuccessfully on the ex-GNR section. The introduction of the 80-class diesel-electric units from 1974 led to further withdrawals, leaving the class represented only by the single unit cars which were retained for reserve and shunting work.

The double-enders were thus the last survivors of the MPD fleet. Nos 63 and 65 were withdrawn in the

late 1970s and No 64 in 1984.

The livery carried by the three single unit MPDs was originally UTA overall unlined Brunswick green, with yellow and black striped warning panels on each end. This livery was later changed to the maroon and silver grey colour scheme the UTA adopted in its final years for ex-NCC section trains, and which NIR adopted as its original livery.

The West Clare diesel railcars

In 1952 CIÉ introduced Walkers diesel railcars to the West Clare, as part of a policy of seeing if a modernised narrow-gauge system could be made economically viable. The four railcars obtained were similar to Nos 19 and 20 introduced on the County Donegal with such success.

The West Clare cars had been ordered in 1951, and the cab sections with the driving bogies were delivered to Inchicore towards the end of that year. They were originally numbered 286 to 289, but by the time they arrived at Ennis they were carrying the numbers 3386 to 3389.

Before being introduced, the standard of the permanent way was improved to allow for faster

CIÉ West Clare section narrow gauge Walker's railcar No 3389 at Ennis.   TJ Edgington, Colour-Rail NG80

running and turntables were lengthened because, like the CDR railcars, the West Clare railcars could only be driven from one end. Two cars operated the main line from Ennis to Kilkee and two operated the branch line from Moyasta Junction to Kilrush. They replaced the normal steam passenger trains and enabled faster services to be provided. A number of new halts were provided to improve passenger access to the railway.

By 1960 the West Clare had become the last surviving Irish narrow gauge line, the CDR having closed in 1959. As a result, no matter what the operating economics may have looked like, the West Clare was a non-standard part of the Irish railway system and CIÉ wanted it closed and replaced by buses. In spite of the considerable traffic carried by the line, CIÉ claimed that it was losing £23,000 per annum.

The transhipment costs of freight, resulting from the change from narrow to broad gauge, did not help when the railway was competing against the more attractive operating costs of the private lorry, nor did the continuing loss of passenger traffic to the private car. Thus, although the railcars and the three diesel locomotives introduced in 1955 had transformed the services and economics of the line, they could not prevent its ultimate closure in 1961.

Car Nos 3388 and 3389, coupled together, were used on track lifting trains. At the time of closure, No 3386 was in Inchicore undergoing overhaul. After closure, at least two of the power bogie/cab units lay in Inchicore until they were scrapped. The coach sections were sold to Bord na Mona where at least one still survives in use.

As with the CDR cars, the West Clare railcars were powered by Gardner 6LW diesel engines. The power train comprised a Don-Flex dry plate clutch, a Meadows four-speed gearbox and from there, through a Hardy-Spicer gearbox, to a Meadows worm gear final drive unit fitted to the rear axle of the power bogie. The two pairs of wheels of the power bogie were connected by coupling rods. Top speed was 38½ mph, in a forward direction only. Maximum speed in reverse was only about seven mph. The power bogie was fitted with a full-fronted cab, like the final CDR railcars, Nos 19 and 20.

The coach sections and trailing bogies were built by CIÉ at Inchicore works. Each complete vehicle weighed about 11 tons and was able to haul a standard passenger coach and a lightweight luggage van. Forty one passengers, all in one class, could be carried and seating was of the standard bus type. Special light-weight passenger trailers, utilising old bus bodies, were constructed in Inchicore to run with the railcars.

Livery was the CIÉ overall green of the 1950s. The fleet number was carried at waist level on the power bogie and at the rear of the coach portion. Strangely the 'Flying Snail' emblem was not carried.

Recent railcar developments

The NIR BREL-Leyland National railbus.

In 1982 NIR introduced a four-wheeled railbus, based on the Leyland National integral-bodied bus design, for use on the Coleraine to Portrush branch. This is a unique vehicle as it was a prototype, produced in 1981 by British Rail Engineering Ltd (BREL) at their Derby works, as part of an attempt to develop the home and overseas market for lightweight diesel units. In this respect, it was similar to the Ganz railcar described above.

In 1978 British Rail's Research and Development Division was re-investigating the possibility of transferring automotive technology to the railways. The first result of this was the production of the Light Experimental Vehicle (LEV) in collaboration with Leyland Vehicles Ltd. This vehicle consisted of two Leyland National bus bodies mounted back to back on a modified four-wheeled underframe built by British Rail Engineering Ltd (BREL). The vehicle was about 36'0" long.

This first experimental unit, LEV 1, operated on trial in the USA, on the Boston to Concord line. The results of tests of the LEV were sufficiently encouraging for two single car pre-production units (R2 and R3) to be built. R3 was tested in revenue service with BR. This is the vehicle which NIR acquired in 1982.

Although the design was not a success as a single unit railbus, failing to attract orders, it developed into the British Rail class 141 lightweight diesel trains. These are two-car units based on the same design elements as R2 and R3. Each car is powered by a single bus-type engine.

When it was running on BR, the railbus carried

NIR BRELL-Leyland railbus R3 near Antrim.

CP Friel

the number RDB977020. Its allocated NIR number is RB3, although it does not carry this on the vehicle.

RB3 is 50'2¼" long by 8'0" wide and weighs 19.4 tons. Power is provided by a Leyland 690 horizontal turbo-charged in-line six-cylinder diesel of 200 hp, mounted beneath the floor. Drive is through a Leyland SCG fully automatic epicyclic gearbox to one axle-mounted flexible drive unit. Maximum speed is 75 mph. Automatic changes are at 16, 30, and 45 mph.

It is fitted with non-standard draw gear, as this is provided only to allow the vehicle to be rescued in the event of a breakdown. As the railbus was designed as a single unit vehicle, and is equipped with a single bus engine, it is incapable of hauling another vehicle.

Seating is open and the vehicle has a capacity of 56 passengers. It has a cab at each end. Its colour scheme is blue and silver, based on the original NIR 'Intercity' livery. This railbus has not been a great success, as it was found to be too small to handle the traffic on the Portrush branch where it was deployed and, to date, no other suitable work has been found for it.

The problems of lack of capacity are illustrated by the following story. When the railbus was operating the Portrush to Coleraine branch, one winter's night at Portrush, a student decided that it was too cold to ride his motorbike to Belfast and tried to travel by rail. Try as they might, he and the conductor couldn't get the motorbike through the doors. Other passengers were getting anxious about their connection at Coleraine, so the conductor reached into his pouch, produced a 50p piece and gave it to the student, saying, "Here, son, here's your money. Now ride this ** thing to Coleraine and get a proper train there!"

It has also been criticised for its poor riding characteristics over jointed track. The continuous 'nodding' motion tends to produce passenger neck ache on a long journey.

At present (2000) the railbus is in reserve and on loan to the Ulster Folk and Transport Museum. However, it has been leased from Translink with a view to being used on the line being built by the Downpatrick Railway Museum in Co Down.

IÉ 2700-class single-unit railcars

In April 1998, IÉ took delivery of the first of a batch of railcars for use on its expanding suburban services (picture on page 47). There are 27 railcars in this batch, built for IÉ in Spain by GEC-Alsthom. They are similar to, and designed to be worked with, the existing 2600-class 'Arrow' railcars built by Mitsui of Japan.

In some ways, the concept of these vehicles is similar to that of the UTA's Multi-Purpose Diesels described above. Like the 2700-class, the MPDs were designed to be worked in multiple and each car had its own power unit, but a number of special double-ended railcars were designed to operate as single-unit cars if required.

The IÉ order is for 25 units. Nos 2701-25 are for making up 12 two-car sets, plus one spare unit. In addition there are two single unit railcars, Nos 2751 and 2752, for use on lightly trafficked services.

These two single unit cars, which were delivered in 1999, and classified MTC, are fitted with a cab at each end. They can accommodate 53 passengers seated and 139 standing and are equipped with a toilet and wheelchair accommodation. If necessary, they can be worked with the other 2600 or 2700-class vehicles.

Each 2700-class car mounts a Cummins NTA 855R1 diesel-hydraulic engine, which provides power to both axles of the leading bogie. This engine is rated at 350 hp at 2,100 rpm and is identical to the engines fitted to the 2600-class Mitsui cars. In addition, each car is fitted with an auxiliary diesel engine to supply power for controls, heating and lighting.

The railcars are designed to run at a maximum speed of 75 mph. Weight is approximately 38.5 tonnes and the length of the MTC single unit cars is 21.59 metres, about one metre longer than the double-unit M1 and M2 cars.

The livery is the same as that carried by the Mitsui 'Arrow' railcars, a variation of the CIÉ black and orange colour scheme.

'Might have been' projects

The DSER enquiries

In 1921, after the failure of a scheme to electrify the line from Dublin to Kingstown, a specification for a petrol-engined railcar was drawn up by the DSER. It was to have the capacity for 50 passengers and be able to complete, in 20 minutes, a six mile run on the level with eight 20 second stops.

Shortly after this specification was produced, the London and North Western Railway offered the DSER a petrol-electric railcar which they had been using in North Wales. Although in good order, the DSER decided it was not suitable for their needs. This car was most likely an experimental car provided by Diamler to the LNWR for test purposes, probably before World War I.

In 1922 quotations were obtained for four Clayton petrol-electric railcars, but nothing further was done to follow up on this. Other enquires were made from a couple of English railway companies, as well as with the MGWR, but again this led nowhere.

The last gasp of this initiative was in 1924 when the American rolling stock manufacturer Brill offered their 'Model 55 Gasoline Rail Car'. However, once again the DSER decided not to proceed down this road.

The Lough Swilly proposal

Following on the success of the early County Donegal railcars, one of which had been used by the CDRJC to tour the Swilly network in 1927, the Londonderry and Lough Swilly Railway decided to investigate the possibility of introducing railcars of their own. In particular, the Swilly were thinking of introducing a railcar onto the Letterkenny-Burtonport line.

In 1931 the Swilly's CME visited the CDR headquarters at Stranorlar to gather information on the CDR railcars. Armstrong-Whitworth were then

London and North Western Railway's Daimler-built petrol-electric railcar, which was offered to the Dublin and South Eastern Railway.
Author's collection

asked to provide a quotation for two railcars. They quoted for two petrol-electric cars at a cost of £2,300 each.

However, the Swilly by that stage had finally committed itself to switching over completely to road transport and no further action was taken on purchasing railcars.

The LMS (NCC) narrow gauge test

In 1933 the Gardner Edwards Diesel Railcar Co supplied a railcar to the Antioquia division of Colombian Railways. This vehicle was bodied by the Service Motor Works of Belfast who also built bus and tram bodies for the Belfast Corporation Transport Department, amongst others.

Before it was shipped to South America, the railcar was tested on the Ballymena to Larne narrow gauge section of the LMS (NCC), but did not seem to have influenced the NCC's policy of operating their narrow gauge lines exclusively with steam traction.

Colombian Railways operated a large number of diesel railcars, but it is not clear whether any further ones were supplied from Belfast.

Chapter 5

Railbuses

It is proposed to deal in this chapter with road buses which were converted for running on the railways. It is not proposed to deal with vehicles, like the BREL-Leyland National railbus, built specifically for use on railways. These are dealt with in the chapter on railcars.

A number of experiments in converting road buses to rail use took place in Great Britain during the early part of the century. In 1906 the Caledonian Railway acquired a Durham-Churchill charabanc and after a period of unsuccessful service it was decided, in 1909, to turn this vehicle into a railbus by the addition of flanged wheels. Towing a four-wheeled luggage wagon, it was used, until 1914, to provide a service between Connel Ferry and Benderloch stations, supplementing steam trains.

In 1922 a new Leyland 32-seat bus, one of a batch of three belonging to the North Eastern Railway, entered service in the York and Selby areas, not as a road bus but as a railbus. Considering the later GNR Howden-Meredith patent for converting road wheels to rail use, it is interesting to note that a photograph of this bus seems to show that its steel rail wheels, with their flanges, are fitted *over* the solid tyres of the road wheels.

Like the Irish railbuses to be described below, these railbuses could operate only on the railway. However, in 1930 the LMS, in England introduced an interesting experimental vehicle called the 'Ro-Railer'. This was a conventional single-deck Karrier bus, with a 26-seater Cravens body, and was converted to run on either road or rail. In the road configuration the railway wheels, set behind the road wheels, were positioned out of the way. On transfer to rail, the steering was locked, the road wheels were raised and the unit became a railbus. This transfer took only a couple of minutes.

The Ro-Railer only lasted in service a few months and this experiment, along with a couple of other similar LMS road/rail hybrid experiments, was abandoned. The reasons for abandonment are not now fully clear, but were probably either lack of passenger capacity or, more likely, lack of interest at the crucial decision-making levels of management.

The Irish standard gauge of 5'3" was especially suited to road to rail conversions, as it was very close to the standard width between road wheels. This made expensive modifications unnecessary. The Irish conversions were not to dual-mode vehicles, like the Ro-Railer, but to purely rail vehicles. To meet growing road competition, they were introduced on routes where passenger traffic was not heavy enough even for a diesel railcar, much less a conventional steam train. They played an important part in keeping branch line services going, particularly in the mid 1930s and, along with railcars, they paved the way for the introduction of multi-unit diesel trains after World War II.

For their operators, railbuses were a low-cost solution to the problem of poorly trafficked lines. From the passenger point of view, they cannot have been great to travel in, with intrusive engine and track noise, and the riding characteristics of a four-wheeled vehicle on fish-plated track were far from smooth.

The NCC railbuses

In 1919 a deputation from Portglenone in Co Antrim asked the Northern Counties Committee of the Midland Railway to provide a motor bus service, to link the railway with the town on a temporary basis until a rail link was planned and put in place. The NCC agreed to do this and began services with a hired charabanc, until the two 45 hp AEC 23-seater buses it had on order were received and put into service. In 1920 traffic was sufficient for a third bus to be added. Later that year, a local firm began running cars in opposition to the buses and this pushed the service into loss. Accordingly, in 1922 the railway bus service was ended and the plan for a

*LMS (NCC) railbus
No 3.*
Author's collection

branch line to the town was dropped.

In 1924, as part of an economy drive by the NCC's new general manager, James Pepper, one of the now redundant AEC petrol-engined road buses, No 3, was converted to run on the railway and was operated on an experimental basis between Coleraine and Portrush for a few months from April 1924.

No 3 ran on road wheels on which the solid tyres had been replaced by railway flanges. Two narrow doors, one on each side, with steps at platform height were fitted at the rear of the bus. The vehicle retained its steering wheel, though one presumes that the front wheels would have been locked in position for railway working. The bus would have had to be turned on a turntable at the end of each trip.

The LMS disposed of the three buses in 1925, the railbus being re-converted to a road bus before being sold to the firm of A Kirk and Co in Belfast.

Finish on this railbus was overall LMS maroon with a white roof. The LMS coat of arms were carried on the rear side panel and the letters NCC, in embellished scroll-work, on the forward side panels. A small number '3'

was carried on the bonnet sides. The bodywork was by Ransome, Sims and Jeffries.

In 1934 and 1936, the LMS had another attempt at the railbus concept. Two 32-seater Leyland Lion road buses were converted for use on the NCC as railbuses.

The two Leylands were part of a batch of five road buses which are recorded as having been transferred from the LMS in England to the NCC. This group of vehicles consisted of three Leyland Lions and two Albion PM28s. All were forward control, half-cab, rear entrance 32-seaters, with LMS bodywork. None of these vehicles appear in the bus

LMS (NCC) railbus No 42. Note the English registration plate and rear view mirror still fitted. Author's collection

fleet lists for the LMS (NCC) and they were supplied to the NCC at a time when their road fleet was about to be taken over by the newly formed NIRTB.

One of the three Leylands was converted into a parcel van, and the two Albions were converted to goods lorries. Further details of this batch of vehicles are given in the road bus section of this publication.

One of the remaining two Leyland Lions, carrying the NCC bus fleet No 42, appears in a photograph, after conversion to a railbus, with its road registration plate CH 7910 still in place. This identifies it as one of a batch of 18 Leyland PLSC3s, built in 1928-29 at the Derby Carriage and Wagon Works for the Sheffield Joint Operators' Committee, in which the LMS had a one-third share. The other shares were held by the local authority and by the London North Eastern Railway. This bus would probably have been used mainly on the Rochdale and Halifax service.

Its LMS fleet number was 19F, 'F' being the LMS classification for buses. It would have originally carried the names of the Joint Operating Committee's participants along the waistband, the LMS coat of arms on the side and the full LMS name along the side boards above the windows. In the NCC photograph these boards are still in place, but are blank.

As railbuses they were fitted with Howden-Meredith patent wheels (see page 108). As with the earlier Midland Railway AEC bus experiment, the Leylands were used on the Coleraine-Portrush line and the Coleraine-Derry line, providing some of the Coleraine-Derry local services introduced in 1936. The timetable allowed them to stop at certain level-crossings to pick up or set down passengers. Tickets were sold by the guard.

Since their capacity was limited, steam trains occasionally had to be substituted and in these cases steps had to be provided for use at intermediate level-crossing stops. Malcolm Speir, the NCC general manager since 1931 and responsible for the introduction of these two railbuses, felt that their use had been a success as a method of building traffic. As with most of the railbuses of this period, they had to be turned on a turntable for the return trip.

Livery was the standard LMS (NCC) bus livery of the period – LMS red below the waist and cream above the waist. The words 'Northern Counties' (in capitals) appeared on the waistband with the fleet number on the front, below the driver's window and rear at waistband level.

It is not now clear when the railbuses were withdrawn from service or what was their ultimate fate.

The CDR railbuses

In 1930 the CDR began to experiment with the use of road buses as feeders to the railway. For this they bought four 36 hp Reo petrol-engined vehicles second-hand from the GNR. These were GNR road bus No 72 and probably also Nos 69, 73, and 75, all of which had been acquired from United Motor Services of Tandragee, when the GNR took that concern over in December 1929. Nos 69 and 72 seated 20 passengers and Nos 73 and 75 each seated 14.

The state of Donegal's roads was so bad that these buses were reduced almost to scrap after a couple of years. In 1933 CDR general manger, Henry Forbes, had the best two converted into railbuses, Nos 9 and 10 in the railcar fleet. No 9 gave good service in this form for a further 16 years, though No 10 was destroyed beyond repair in a fire at Ballyshannon in 1939.

The livery carried by the railbuses was that of the railcar fleet and passenger coaches, geranium red below the waist and cream above, with the CDR coat of arms on the side.

No further railbuses were added to the fleet, as the CDR decided to concentrate on a policy of dieselisation, based on purpose-built railcars for the railway's passenger traffic.

The GNR railbuses

With the experience of building the CDR railcars in the 1920s and 1930s, the GNR decided to introduce similar vehicles for its own use to challenge the increasing road competition which was hitting traffic on minor lines. As a result the GNR, probably more than any other railway company in the British Isles, pioneered the use of diesel traction on lines where passenger traffic was light.

The first vehicles to emerge as a result of this policy were the railcars A and B, described earlier (page 88), but it was soon clear that something smaller than railcars would be needed if economic provision was to be made on very lightly trafficked lines. The solution decided upon was to convert

GNR railbus F. Detail of Howden-Meredith wheels. Duffners

pneumatic tyre, which was of a size suitable to use with scrap steel tyres from locomotive bogies. The insides of the steel tyres were machined to fit correctly over the rubber tyres.

Five buses were so converted for working on the GNR, in addition to two supplied for use on the DNGR and three for the SLNCR. All but one of these buses had been acquired by the GNR from private companies they bought out between 1929 and 1931.

Railbuses, like the GNR railcars which came out during the same period, initially carried letters in the railcar series, rather than numbers. Later in their lives numbers were substituted for the letters.

The first railbus was lettered D and it went into service in 1934. This railbus had a chassis which dated from 1926, having been GNR AEC 413 road bus No 5. It was fitted with a 30-seater body dating from 1927, from road bus No 186, an ADC 416. When railcar D appeared, the railbus was re-designated 'D1'. It worked between Dundalk and Clones and was sold to the SLNCR in 1939.

Railbus E followed later in 1934. It was converted from 31-seater road bus No 7, an AEC 426, built in 1928. It was used on the Scarva-Banbridge branch, making it possible for this to be re-opened to passenger traffic. It was also loaned for a period to the SLNCR to allow that company to

some road buses to rail use, the first of these appearing in 1934.

These GNR road bus conversions were fitted with road wheels converted for rail use, the Howden-Meredith patented wheel. This wheel was the result of collaboration between two Dundalk engineers and the Dunlop Rubber Co.

The patent wheels used in Ireland were different from the Michelin tyres used on experimental railcars introduced by the LMS during the same period. On the LMS vehicles the rubber road tyre ran on the rail, with a steel rail flange behind it. The Howden-Meredith wheel had a steel tyre, with the flange fitted outside the rubber tyre. The inner surface of the steel rail tyre was designed to allow the road wheel to fit tightly inside.

This design made puncturing almost impossible and also eliminated tread wear. To safeguard against the risk of accidental tyre deflation, a safety brake block was fitted above the wheel. In the event of a rubber tyre deflation, it applied the brake, held the steel rim vertical and switched off the engine. A special offset inner steel wheel was made by Dunlop to take the

GNR railbus E in original form, on trial at Manorhamilton on the SLNCR. With a new body this vehicle became No 1 and is now in the UFTM.
HC Casserley.

assess the possible usefulness of this type of vehicle on their line. It was redesignated 'E2' in 1936 when railcar E went into traffic and became No 1 in 1947.

During its life, it was rebodied with a standard half-cab GNR body, and its petrol engine replaced with a Gardner diesel engine. In 1956 it was converted to a civil engineer's vehicle and re-numbered 8178. At this time, it was also fitted with a reverse gearbox so that it could run equally well in either direction.

It was transferred to UTA ownership in 1958 where it was used by the Civil Engineer's department. In 1963 it had a puncture, after which it lay out of service in Goraghwood until a spare wheel was located in Dundalk. It was finally withdrawn from service in 1965 and was transferred to the Belfast Transport Museum's collection. It is now on display at the UFTM at Cultra, Co Down.

Railbus F, re-designated 'F3' in 1938 when railcar F went into service, appeared in 1935. This was a conversion from 30-seater road bus No 6, an AEC 426. It was very badly damaged at Dundalk in 1944 and was replaced in the same year by another vehicle, also designated F3. It became No 2 in 1947 and went into CIÉ stock in 1958.

In its final form, railbus No 1 was entered by a folding door opening straight off the rear platform, whilst No 2 had a small porch protecting the rear entrance.

Two DNGR railbuses were taken into GNR stock in 1948 and renumbered 3 and 4. No 3 was scrapped in 1955, but No 4 was transferred to the Permanent Way department and renumbered 8177 in 1956. It was transferred to CIÉ in 1958.

The second GNR railbus F3 and the two DNGR vehicles, seem to have been converted from GNR road buses Nos 33, 34 and 35 – all ADC 416s acquired on the absorption of Fairways Ltd of Dublin. However, from the records it is not now clear which bus became which railbus.

GNR railbuses D1, E2, the first F3 and SLNCR railbus A were all converted from buses originally belonging to the Louth and Meath Omnibus Co, taken over by the GNR in February 1929.

Although fitted with petrol engines when built, by the time of their conversion into railbuses, all these vehicles had been fitted with Gardner 4LW diesel engines. The exception was SLNCR railbus A,

where the conversion was made in 1936, a year after it had entered service.

On the original conversions, Howden-Meredith wheels were fitted front and back. When the vehicles were found not to be operating the track circuits, they were re-fitted on the front with solid wheels and a solid connecting wagon axle. This reduced the passenger comfort of the vehicles. The solid wheels did not properly solve the track circuiting problem so signalmen were issued with special instructions for handling the passage of railbuses.

Certain of the railbuses had been built originally with full-front cabs, but to simplify engine access they were later converted to half-cab layout. In their road bus form, passenger entrances were at the front but, when converted, it was found that the best passenger access layout was for a station platform-height rear platform to be fitted, with trailing steps at either side. This allowed boarding both at conventional station platforms and ground-level intermediate stops such as level-crossings.

This conversion was not applied to railbus D, the prototype GNR railbus, which went to the Sligo and Leitrim in 1939. It had two passenger entrances located, one on each side, immediately behind the drivers cab.

One of the GNR railbuses was tried out on the Hill of Howth tramway in the early 1930s, with a view to replacing the trams. However, on the day the bus was tried out, it jumped the rails no less than six times on its way up from Howth to the summit of the Hill of Howth. Worse was to follow. On the descent, the bus ran away on its driver, hit the points at the Howth Demesne loop and ended up on its side. Fortunately, nobody was badly hurt. That ended the experiment with railbuses on the Howth line for good.

The livery was the same as that applied to the railcars – GNR Oxford blue below the waist and cream above it, with the GNR coat of arms applied to the sides.

The DNGR railbuses

The DNGR was a small outpost of the London and North Western Railway, linking the ferry port of Greenore with both Dundalk and Newry on the GNR system. As Greenore never grew to rival Larne, Belfast or Kingstown (Dun Laoghaire) the railway

Ex-Dundalk, Newry and Greenore railbus No 2 as GNR railbus No 4 at Dundalk.   Author's collection

was never very prosperous, but it survived into LMS and finally British Railways ownership before being closed to traffic in 1952 and wound up in 1957.

After the partition of Ireland in 1921, the DNGR found part of the northern leg of its system, from the border to Newry, in Northern Ireland and the rest of the railway in the Irish Free State. As with the GNR this made it difficult to absorb the line into the national railway systems of either state.

After a period of negotiations, agreement was reached between the LMS and the GNR whereby the latter would work the railway on behalf of the LMS, thus allowing greater economies of operation. This arrangement came into force in 1932-33.

Early in 1935, the GNR tried out one of its railbuses on the Newry-Greenore service on the three days of the week when there was no steamer sailing and therefore no need for vans to convey parcels, goods or livestock. A photograph would seem to indicate that this was GNR railbus F.

The experiment was successful enough to warrant its extension to the Greenore-Dundalk section and a second bus was introduced in 1935 for this purpose. As a result of the introduction of the railbuses, four new halts were established on the Dundalk-Greenore line, between the stations The Bush and Greenore, and two railbus 'request stops' on the Newry-Greenore section.

As noted above, in the section on GNR railbuses, the two buses were transferred into the ownership of the DNGR and were given the numbers 1 and 2. They were converted to run on gas during World War II.

Although the railbuses were instrumental in developing passenger traffic in the area, they never became a paying proposition and were eventually sold back to the GNR in 1947.

The SLNCR railbuses

In 1933 a private bus company, owned by a Mr Appleby, began operating a through service between Enniskillen and Sligo, the two ends of the SLNCR system. The SLNCR put out feelers to buy the service from Appleby, but without success. In 1935 the NIRTB was established and in 1937 Mr Appleby's service was truncated to a Blacklion-Sligo service, the Belcoo-Enniskillen section coming under the control of the NIRTB (Belcoo and Blacklion are twin villages straddling the Irish border).

In order to meet the challenge of this road competition, as well as exploring the possibility of buying it out, the SLNCR were watching with interest the GNR's experiments with railcars. In 1932 one of these was tried out on the SLNCR. Its management were impressed with the railcar's low operating costs, its ability to pick up and set down passengers almost anywhere, and to negotiate their line's gradients and curves.

A decision to buy a light, self-propelled passenger vehicle was taken in 1934, by which time the GNR railbuses had appeared. Ex-GNR road bus No 4 was bought from the GNR and was converted into a railbus for the SLNCR at Dundalk. It was delivered in 1935, along with a specially built two-ton capacity low-sided, tarpaulin covered luggage trailer to run with it.

In GNR-style, this first railbus was designated railbus A. This was an AEC Type 413 road bus, fitted with an AEC Reliance six-cylinder 40 hp petrol engine. It was built originally in 1926 for the Louth and Meath Omnibus Co.

On conversion, it was equipped with Howden-Meredith patent wheels. At the rear, a platform with trailing steps was added on each side, giving the 7'0" wide bus an overall width of 9'0". This was necessary to allow passengers to alight or join at stations, while the steps allowed access or egress at intermediate points, such as level-crossings. The bodywork was made narrower at the rea so that the outwards-opening swing doors were within gauge when opened. Seating capacity was 32.

In 1936 GNR railbus E, which was fitted with a diesel engine, was tested on the line. As a result, railbus A was re-fitted with a Gardner 4LW diesel engine, which improved fuel consumption from nine mpg to 21 mpg. Railbus A was broken up in 1939 after being involved in a collision with a locomotive.

In 1936 the decision was taken to order a second railbus and in 1938 No 2A was delivered. This was an ex-GNR ADC 416 road bus, No 185, but had been equipped by the GNR with a Gardner 4LW diesel. The conversion to a railbus involved fitting a totally new body and Howden-Meredith wheels. '2A' ran with the luggage trailer of the ill-fated railbus A.

A replacement for the destroyed railbus A was obtained in the form of the prototype GNR railbus D1 in 1939. This railbus had a chassis which dated from 1926, having been GNR road bus No 5, but a body from 1927, from road bus No 186. It differed from the other SLNCR railbuses in that it had inward-folding doors on each side just behind the driver's cab, rather than the rear doors and platforms. It was designated A, the letter of the bus it replaced, and incorporated the diesel engine from the original railbus A.

In 1942 it was decided to provide a luggage trailer for railbus A. A two-ton vehicle was constructed out of the bogie frame and wheels of a Great Southern Railways Sentinel steam railcar and a trailer body built by the SLNCR at their Manorhamilton workshops. This new trailer was almost as high as the railbus itself.

In 1950 railbus A went for overhaul to Dundalk, where it was found that the body was in very poor condition. It was decided to replace it with a new body from withdrawn GNR road bus No 310. This conversion left the bus looking somewhat similar to 2A, with rear platform and steps. No 2A went for overhaul in 1954, but subsequently the GNR made it clear that they were no longer in a position to supply some of the spare parts required.

Incidentally, GNR road bus No 310, which provided the body for the second SLNCR railbus A mentioned above, had appeared in a GNR publicity photograph as a railbus in 1934, when new. However, it never actually ran as a railbus at that time.

The SLNCR railbuses survived to the closure of the railway in 1957. Railbuses and trailers were auctioned in 1958 and 1959 and were broken up soon afterwards.

The livery used on the Sligo and Leitrim railbuses was two-tone green, with white roofs. Below the waist was Regent green, with olive green above, separated by one or two narrow black lines.

The CIÉ railbus programme

The last road bus to be converted to a railbus in Ireland was fleet No A8, an AEC Regal IV, 34-seater single-decker, built in 1934 for the Dublin United Tramway Co. This bus was withdrawn from service in 1951 and converted into a railbus by CIÉ in 1953.

Having been given the railcar fleet number 2508, it was used on the Thurles-Clonmel line for about two years from early 1954, but often proved inadequate for the traffic available. Having lain out of use for a number of years, it was finally officially withdrawn for scrapping in early 1962 (page 114).

There is some evidence that CIÉ intended to carry out a number of such conversions for use on lightly trafficked branch lines. This plan was said to be the brainchild of GB Howden, who had been appointed general manager of CIÉ in 1950. Howden had previously been general manager and Chief Engineer of the GNR and so had been involved in their railbus programme. When Howden left CIÉ in 1953, to become chairman of the UTA, the CIÉ railbus programme seems to have been abandoned.

Railbus 2508 carried the two-tone green CIÉ livery on the period, as applied to the road bus fleet.

CIÉ narrow gauge bus-coaches

As well as A8, which became railbus 2508 mentioned above, a second bus is identified in the CIÉ fleet lists as having been withdrawn for conversion to an unidentified 'railcar'. This was fleet No TP17, a 32-seater Leyland TS6 coach built by the GSR in 1935 and withdrawn from service in 1951. There is no evidence that this bus ever entered service as a railbus.

However, in 1951, when CIÉ decided to introduce diesel railcars to the West Clare system, three ex-Tralee and Dingle coaches were converted at Inchicore to operate with the new Walkers railcars. The conversion took the form of mounting old road bus bodies on to the Tralee and Dingle underframes. The new coaches were fitted with bus seats and could carry 40 passengers. It is known that six ex-GSR buses provided the bodies for these conversions, two bus bodies to each coach.

The West Clare coaches were numbered 46C-48C and were painted in the CIÉ green livery of the period. One has survived into preservation as part of the Cavan and Leitrim Railway's collection at Dromod.

A similar coach, incorporating two ex-GSR bus bodies, was constructed on the frames of Cavan and Leitrim bogie coach No 7. This coach was fitted with bus seats and electric bus lights, operated off batteries. It was put into service, in this form, in 1953 and on closure of the C&L was disposed of to Bord na Mona.

It is possible that TP17 was used in one of these bus-coach conversions. The CIÉ omnibus fleet lists are silent on this point.

GNR railbus No 1 at Dunmurry on 21 March 1993, on its way to the Ulster Folk and Transport Museum at Cultra.
Author

SLNCR railbus 2A on Sligo turntable in 1956. Note the specially built luggage van and the Howden-Meredith wheels all round.
TB Owen, Colour-Rail IR187

CIÉ railbus 2508 out of service at Inchicore, in 1961. This was an ex-DUTC AEC Regal IV.

Colour-Rail IR405

CIÉ Drewry inspection car No 2 at Kildare in June 1961. This was one of four similar vehicles, built in 1926-27.

Colour-Rail IR296

Preserved CDR railcar No 12 at the Foyle Valley Railway museum, with No 18 behind.   Author

GNR Railcar A at Drogheda in April 1956. It is preparing to work the 3.40 pm service to Oldcastle.
TJ Edgington, Colour-Rail IR130

GNR railcar C1 at Cavan, in 1956.

C Banks collection, Colour-Rail IR281

Ex-GNR railcar F, as UTA No 104, at Goraghwood in 1963. Note railbus No 8178 (ex No 1) on the right, with a punctured tyre (see page 111).

JG Dewing, Colour-Rail IR306

LMS (NCC) railcar No 4 in UTA livery at York Road, Belfast. Author's collection

SLNCR railcar B at Manorhamilton, in 1956. The powered end is furtherest from the camera. The half-cab driving position can be clearly seen. C Banks collection, Colour-Rail IR282

CDR railcar No 19, with trailer No 3, at Stranorlar in 1958.   JM Bairstow, Colour-Rail NG70

CIÉ West Clare section railcars on track lifting duties at Ennis, in 1961.   TJ Edgington, Colour-Rail NG81

NIR railbus at Portrush, in 1981.

N Johnston

Return to Great Victoria Street station, Belfast. 80-class railcars Nos 96 and 98 on the first day of service to the re-opened station in 1995.

Author

NIR 'Castle' class railcar in new Translink colours at Carnlea in 1999.   N Johnston

Postcard of Portrush LMS (NCC) railway station, with NCC buses in the foreground. A Giant's Causeway tram is just visible to the left of the station.
Author's collection

CDR Leyland Tiger Cub bus line up. Five of the fleet of six are shown here.   JA Curran

LLSR Leyland Atlantean bus under repair in the old locomotive workshop, Pennyburn, Derry. Three foot gauge track can be clearly seen on the left.   Author

LLSR No 149. This vehicle was originally UTA Leyland Leopard No U 482, one of the 'Wolfhound' fleet introduced in 1965.
Photobus

GNR-Gardner bus No 389, with Park Royal bodywork, built in 1951. It is shown as preserved by the Transport Museum Society of Ireland.
Author

GNR AEC Regent III double-decker bus as CIÉ AR295. This bus had Park Royal bodywork and was new in 1948.
Photobus

CIÉ E-class single-decker. This Leyland L2 Leopard was introduced in 1961. Some of this type were hired to the CDR .
Author

Three generations of CIÉ double-deckers: Bombardier, Van Hool-McArdle and Leyland.   Photobus

An Irizar City-Century Scania coach, belonging to Bus Éireann. This vehicle, which was new in 1999, is pictured at the Irish Transport Trust rally of that year.

Author

Wright bodied Volvo B6BLE at Poolbeg Street on a Dublin 'City Imp' service in 1999, shortly after entering service.
Will Hughes

Artist's impression of the proposed electric tramcars to be built for the LUAS light rail system in Dublin.
LRT Information Office

Preserved 1947-built Leyland Tiger PS1 No A 515, in UTA livery, at the 1999 Irish Transport Trust rally. This vehicle in now owned by Ulsterbus.
Author

Citybus Daimler Roadliner on Citylink service at Central Station Belfast. This ex-Belfast Corporation bus was one of a motley collection used on this service.
Photobus

NIR Rail-Link bus. Citybus Tiger with Alexanders Q-type body, in the now defunct NIR 'Suburban sector'
colours.
Author

NIR Inter-City liveried Leyland Tiger Van Hool Alize coach, at Downpatrick Railway Museum.
Author

The return of the 'green bus'. The new Translink Citybus livery on a Volvo B10 with Alexander's body at Donegall Square West, Belfast, in June 1999. Author

Volvo B10 with Wright's body, carrying the Ulsterbus version of the new Translink livery. This example is at Queen Street, Belfast, in February 2000. Author

Chapter 6

Road buses

In this chapter, it is intended to outline the development of railway involvement in the provision of road passenger services. Where the number of vehicles is small, details will be included in the text, but where large fleets were operated, space permits only a general sampling of the main trends and examples of vehicles operated.

The use of road buses by railway operators described in this section can be divided into three broad phases. Prior to World War I, some railway companies obtained limited powers to operate road services. The BNCR obtained such powers in 1899 and the MGWR and GSWR in 1903. Some of these companies operated a few routes using buses – horse, steam, or motor-powered – as feeders to the railways. In many areas, the biggest problem faced was the poor state of the roads, especially in winter. On occasions, the local authorities would even bill the railway companies for damage done by their vehicles to the unmetaled roads.

The second phase developed after World War I when a flood of cheap surplus army motor vehicles became available, along with the demobilised men with the skills needed to drive and maintain them. These vehicles, the products of a motor vehicle industry greatly expanded during the war, could be bought under hire purchase schemes. This put them within the reach of many potential entrepreneurs.

There were few regulations governing the use of such road vehicles. On the other hand, the railways were heavily regulated as to operating practices, safety, payment and conditions of staff, and had to pay for the upkeep, signalling and policing of the permanent way.

Another major problem for the railway companies, in their traditional role as common carriers, was that they were controlled by law, as to the rates and fares they could charge, and had to accept all traffic offered for carriage by rail. Their road competitors were not so encumbered.

A further feature of this period was the competition on fares. Both road and rail operators cut fares to attract and hold custom. This had the effects of forcing the weaker bus operators out of business and of badly damaging the railways, financially, at a time when operating costs were rising. For example, during 1927 the NCC's steep fare cuts increased traffic by 78%, but revenues by only 16%.

The easy availability of road vehicles during this post war period, plus the improvements in the surfacing of roads, led to growing competition with the railways. As has been outlined in previous chapters, the railways responded in a number of ways, one of which was the development of rail vehicles which could be operated at lower costs than traditional steam trains. However, they also began to lobby government for the right to run road services themselves and later, when this right had been acquired, for the monopoly of such services in their areas. As a result of these changes, in many instances the railway companies bought out competing road competitors and in several cases they ended up as substantial bus operators in their own right.

A contemporary academic view, by JC Conroy in his 1928 MA thesis, is interesting in that it sets out many arguments which have been discussed right down to the present. He noted that the railways had lost much of the short haul transport to the roads, but that they had also gained indirectly from the roads. The principal points he made were that the availability of cheap motorised transport generated an increase in tourist traffic; the existence of the motor lorry allowed traffic to be tapped into by the railways from a much wider catchment area; and new types of traffic which rail could carry were on the increase, eg petrol, road building materials, etc. Therefore, the traffic lost to the roads was partly counter-balanced by the new business created by the development of better communications.

Nonetheless, even in the 1920s Conroy

recognised that the heavy lorry did not pay its way towards the roads it used. In his opinion, unrestricted road transport was not an economical proposition for the new Irish Free State. His view was that road transport should complement rail, and not be viewed as a substitute, but that the Irish railway network should be rationalised. Road transport should, however, be subjected to restrictions similar to those imposed on the railways in the Irish Free State by the 1927 Railways (Road Motor Services) Act.

The third phase developed just before World War II and reached its peak in the 1950s and 1960s. During this period, road competition had become so intense that it threatened the very existence of the railways in many parts of the country. The solution to this problem took a number of forms. In a couple of cases, minor railway companies switched totally to road operations by choice. A more important trend, however, was for road and rail services providers to be merged by government legislation in order to provide an integrated service under public ownership. Nonetheless, in all cases the net result was a reduction in the route mileage of railways all over the country.

Since the late 1960s, the fashion has been to split the nationalised, integrated road and rail operations and place them under the control of semi-independent, competing companies under the general control of a holding company. Ownership of the holding company continued to be vested in the state.

This happened in Northern Ireland when the state-owned integrated transport operator, the Ulster Transport Authority (UTA), was split up into three semi-state companies – Northern Ireland Railways Ltd, Ulsterbus Ltd and Northern Ireland Carriers Ltd – under the general direction of the state-owned Northern Ireland Transport Holding Company.

Twenty years later, in 1987, the integrated state transport company in the Republic of Ireland, CIÉ, was also split up into three main subsidiaries. These are Iarnród Éireann (Irish Rail), Bus Éireann (Irish Bus), and Bus Átha Cliath (Dublin Bus), under the general direction of a holding company, a re-formed CIÉ

In January 1995, in a return to the integrated model, the British government announced that, in Northern Ireland, the two bus companies, Ulsterbus Ltd and Citybus Ltd, would be merged with the railway company Northern Ireland Railways Co Ltd. This created a single operating entity under the control of the Northern Ireland Transport Holding Company. When this took place, under the brand name 'Translink', the organisation of public transport in the province returned substantially to the position it had been in between 1948 and 1967.

Early experiments with road services

The Dublin and Kingstown Railway

The practice of co-operation between road and rail services dates from the beginnings of railways in Ireland. The promoters of the Dublin and Kingstown Railway arranged with private operators that 12 cars would connect with the trains at Kingstown. However, this agreement was not a success and was allowed to lapse.

The promoters of the DKR also planned to have an omnibus service connecting with the trains at their Westland Row terminus in Dublin. They entered into an agreement with a hotelier who had experience with omnibus operations. He started to operate a service between his hotel, close to the city centre, and Westland Row. However, in 1835 the city's hackney car owners took legal action against the railway and succeeded in having this service stopped.

The Dublin and Kingstown also arranged for omnibus connections to Bray from their Kingstown terminus. One of these services was operated by a Bray hotelier. The railway provided him with a bus which he then operated to connect with certain trains. Bus tickets could be purchased at the Westland Row station for this service and for those of other operators who had similar contracts with the railway company.

These services to Bray were regarded by the DKR as extensions to the railway and attracted many additional passengers to the trains. Parcels as well as passengers were carried.

The MGWR

From the earliest days the MGWR was involved in subsidising road services connecting with the railway services at various points on its network.

Shortly after the opening of the MGWR line to Galway, in 1851, the 'Connemara Tour' became very

MGWR Karrier 14-seater charabanc.   Author's collection

popular. The tourist had to travel from Galway to Clifden by 'long car' and from there proceeded to Westport. The cars operated from June to September and they started from the hotels in both Clifden and Westport. The tours, by 1894, were operated by a Mr McKeown of Leenane, with a subsidy from the railway company.

The MGWR was anxious to encourage the tourist traffic and, long before they had extended their own line to Clifden, they were issuing special inclusive tickets. After the extension to Clifden, in 1895, the company built a new hotel at Recess overlooking Glendalough Lake (in 1898). The railway continued to work with Mr McKeown in promoting the long car tourist service until 1910, when it began to express dissatisfaction with his provision.

In 1903 the Midland Great Western Railway Act obtained the powers to operate road services between their hotels and places of interest. They were also granted powers to:

> ... supply coaches or motors for the conveyance of parcels, passengers or goods in connection with, or extension of, their railway system

These very limited powers authorised the company to run buses only for specified purposes over specific roads. The MGWR's powers were the widest of any of the railway companies, but even here there was the restriction that their buses must start from one of the company's railway stations.

In 1910 the MGWR decided to operate the

Clifden-Westport service itself. In order to do this it ordered from Archers of Dublin, in 1911, three 'Windermere' charabancs built by Commercial Cars (from 1926 to be called Commer). According to company brochures, these had a capacity of 15 first-class passengers. However, they were underpowered and so not very successful. In 1914 an additional car was ordered from Chambers of Belfast, while in 1915 another car was ordered for the Recess hotel.

While the cars were employed mainly on the Connemara Tour, they also made trips to and from the MGWR hotels at Mallaranny and Recess. This coach service suffered from the effects of World War I and the vehicles were sold in 1918.

In 1924, only months before the MGWR was amalgamated into the GSR, it was decided to order three 14-seater Karrier cars for the Clifden to Mallaranny service. It is not clear what model of Karrier these were, but if they were like the Commercial Cars buses, they would probably have been the Karrier 'Z' type, a one ton 14-seater, introduced in 1924.

The MGWR also had an agreement with a Mr Hooey, during 1907-08, for a bus service between their Broadstone terminus in Dublin and the city centre. In 1911 meetings were held with the Dublin United Tramway Company for a possible tram service between the same two points, but nothing came of this.

The MGWR buses were based on the 'Norfolk convertible country house and estate car' built by Commercial Cars. This could be either an omnibus, a shooting brake or an estate lorry. For use as a bus, it could carry up to 15 passengers and about a ton of luggage. The MGWR buses were roofed, open sided, with drop canvas screens for protection against the weather. They were fitted with a roof luggage rack and a seat on the running board for a courier. The vehicles were chain driven.

The Commercial Cars buses were painted green with gold lettering. They were lined out, with the company's coat of arms displayed on the centre of the waist panel with the words 'Midland, Great Western Railway of Ireland Company' behind the crest and the back of the vehicle. The letters 'MGWR' were displayed on the rear of the buses on the panel below the back windows, with the words 'Tourist Car' in small lettering and 'Clifden-Westport' in larger lettering on the panel below that. The Karrier buses were painted brown, the colour of the railway's passenger carriages at the time.

The Commercial Cars bus registration numbers were the Galway registrations IM 179-181. However, a posed shot, taken of one of the buses outside Broadstone, shows it carrying the Bedfordshire registration trade plate BM 9 B. The Commercial Cars factory was in Luton, Bedfordshire, so it is probable that this photo was taken when the bus was just delivered from England.

Details of the Chambers cars are not clear, but the only suitable chassis that they produced from 1911 until after World War I, was the 12-16 hp model which had a vertical 4 cylinder engine. To this chassis could be added one of about ten standard bodies built by Chambers. They could also build one-off bodies to order. However, the 12-16 hp was not a PSV chassis and it is therefore likely that the Chambers cars were fitted either with a tourer body or a customised bus-type body, similar to that fitted to the Chambers ambulance supplied to the Belfast Guardians. This had a capacity of seven adults in addition to the driver.

The BCDR

Because of the gaps in the coverage of its natural hinterland by its rail network, the BCDR became involved in the road transport business at an early stage. A timetable dated 1 May 1854, for the Belfast to Holywood line, the terminus of the branch at that time, indicates that a horse omnibus service connecting Holywood and Bangor was available:

> An Omnibus connexion with the Trains leaves Bangor every week-day morning, at a quarter before 10, to arrive in Holywood in time for the 11 Train to Belfast; and returns to Holywood on the arrival of the 6-30 Train.

> An Omnibus leaves Holywood every week-day, morning, on the arrival of the 10-30 Train, and returns from Bangor in the evening, at ¼ before 9, to arrive in Holywood in time for the 10 train to Belfast.

The fares on this service were Belfast to Bangor (train and bus inclusive), 1s 0d and from Bangor to Holywood, or the reverse, 8d. It is not clear whether these buses were owned by the BCDR or operated on its behalf by a private provider, but it is clear that the service was provided as an extension to the rail service.

In 1904 a deputation from the BCDR went to England to inspect the road motor services being operated by Eastbourne Corporation, and by the GWR, with a view to operating a motor bus service between Bangor and Donaghadee. However, in 1905 they decided that motor buses were not a commercial success and decided against setting up this service. Why they should have come to this decision is not clear, as the services of both the English operators were successful.

In the period from just before World War I, until the mid 1920s, the BCDR arranged many car services with independent operators. These services were operated as feeders to the railway. Up until World War I, the services were operated by horse-drawn vehicles but, by 1914, quite a few operators had changed over to mechanically propelled vehicles of one sort or another. In this way, through travel to Belfast was encouraged from points off the main rail network.

In May 1916 Norton of Kilkeel placed a 16-seat Selden motor bus into service between Warrenpoint and Kilkeel to replace their horse-drawn 'long cars', which they had been using for the service between Newcastle and Warrenpoint. Shortly after this the BCDR decided to start operating its own buses.

In August 1916 the BCDR introduced their own bus service between Newcastle and Kilkeel, operated by a covered Dennis motor bus acquired from Charles

*Belfast and Northern Counties Railway
steam bus, one of two used on the
Whiteabbey to Greenisland service.*
Author's collection

Hurst of Belfast. This bus was stabled at Kilkeel.

The BCDR had intended to buy three buses, but were limited to one by wartime restrictions. In September, because of the difficulty in getting petrol, Norton of Kilkeel, who had introduced his Selden motor bus that year, had to re-introduce his long cars and sold the Selden to the BCDR. It joined the Dennis on the Newcastle-Kilkeel run.

In 1919 the Selden was sold and replaced with a new Dennis chassis from Hurst, which was bodied by the BCDR at their Queen's Quay workshops. The original 1916 Dennis lasted until 1923. In the inter-war years, the Newcastle-Kilkeel service would be operated by Leyland PLSC3s, connecting with the trains at Newcastle. Like Reo, Selden was an American manufacturer, based in New York state. Given that this was an early example, the BCDR bus would have been chain driven.

The BNCR

In 1899, an Act of Parliament was passed for the BNCR which, among other things, gave the company the right to:

> ... provide, own, work and use coaches motor cars and other vehicles to be drawn or moved by animal power, electricity or any mechanical power for the conveyance of passengers ... in connection with or in the extension of their railway system ...

In the same year, the company introduced a Milnes-Daimler 6 hp petrol van to be used for parcel deliveries in the Ballymena area, though it was later moved to Belfast. This van may have been fitted with passenger seats, but this is not now known for sure.

There is a possibility that the Milnes-Daimler vehicle was a conversion to a small bus. This company was amongst the best known of the early bus builders. The company was formed from an association between the British tramcar builder Milnes, and the German company Daimler. The first proper bus it built was in 1902 for Portsmouth and Gosport Motors.

The BNCR had horse buses operating as feeders to Whiteabbey and Greenisland stations before the passage of the 1899 Act. In 1902 these were replaced by two 24 hp Thornycroft steam buses, each capable of carrying 14 passengers. These Thornycrofts had the distinction of being the first railway-owned, mechanised buses in the British Isles. A similar, but roofed, double-decker bus was introduced in London in the same year, but operated there for only a brief period.

The first bus was tried out in Belfast for a short period, before being transferred to the Whiteabbey-Greenisland service. The second bus had its trials in London, before being shipped over to Ulster to work in the same area. The BNCR buses ran for a least six years, before being replaced by horse buses. The steam bus chassis were converted to lorries, one of which worked until 1925. The Whiteabbey-Greenisland service was successful in attracting people to use the trains and was operated until 1925, when it was discontinued by the LMS (NCC).

An undated, but obviously turn of the century, photograph also exists of an open-top, double-deck, horse bus displaying the BNCR coat of arms. The photograph carries the caption: 'The Old Railway Bus, Eglinton', so it would seem that the BNCR operated bus connections on that part of their system, as well as in the Belfast area.

Also in 1902, the chairman of the NCC announced that the company had been requested by representatives from a number of areas to provide branch line connections. As traffic would not warrant building such lines, it was felt that road transport might meet their needs. As a result, a Thornycroft steam wagon was introduced in the Cookstown area in 1902. However, it had to be withdrawn by the MR (NCC) in 1906 because of complaints, and a claim for damages from the Londonderry and Antrim County Councils as a result of the damage it was doing to the roads.

The introduction of this steam lorry was significant for several reasons. Firstly, it indicated that further extensions to the rail network could not be justified, given the low level of potential traffic. Secondly, it indicated a realisation by the NCC management at that time that further extensions to their system could be made by using road transport.

Thus, the MR (NCC) directors seem to have realised, at this early stage, that they were not in the railway business as such, but in the mass transit of people and freight business and that the transport mode selected should suit the conditions of traffic in the area. This was an important concept which re-emerged in the mind of the LMS (NCC) general manager, Malcolm Speir, a number of years later.

The MR (NCC)

The Midland Railway of England absorbed the BNCR in 1903 and, in 1904, a Milnes-Daimler 18 hp 16-seat vehicle was brought from England and taken on a trial run from Larne to Portrush along the coast road. Nothing more seems to have come of this experiment, but in 1905 two Thornycroft petrol-engined motor charabancs were bought for coast road tours work. One was used between the railway's Northern Counties Hotel in Portrush and the Giant's Causeway and the second provided a service between Parkmore station, on the NCC narrow-gauge, at the head of Glenariff, to Cushendall, on the coast. For winter work it was possible to fit this vehicle with a fully enclosed body, complete with railway carriage torpedo ventilators on the roof. A posed publicity shot of one of these buses taken at Fortwilliam in Belfast (below), shows it carrying the registration plate of one of the earlier steam buses.

The poor state of the roads led to the NCC withdrawing their Parkmore-Cushendall service in

MR (NCC) Thorneycroft charabanc No 1, carrying a steam bus registration plate, in a posed shot at Fortwilliam Park, Belfast. It is carrying its open summer body, but in the winter had enclosed boydwork.

WH Montgomery collection

1907, after which they ran tours using both charabancs from Portrush. This lasted until 1915, when the vehicles were sold to Henry McNeill of Larne. After the NCC withdrew their Cushendall service in 1907, McNeill provided a replacment service, subsidised by the NCC. He was contracted initially to provide a horse-drawn long car for the mail, three horse-charabancs for summer tourist traffic and a closed omnibus for winter services. McNeill continued to work the Cushendall route until 1928, when he was bought out by the LMS (NCC) as part of their expansion into road operations.

In 1912, the provision of a bus service between York Road station and the cross-channel steamer terminals at nearby Donegall Quay was considered. It was probably intended to be operated by a horse bus, although it is not clear whether it was ever introduced. However, an earlier picture of York Road station, showing the Belfast Street Tramway Co's horse trams (replaced by electric trams in 1905) going into, and coming out of, the station's tram bay, also shows a horse omnibus passing the front of the Midland Hotel, from the direction of the steamer berths. It is not clear from the print who operated this horse bus, but it is known that, at that time, the BNCR, as well as Belfast's prestigious Grand Central Hotel and the Belfast Street Tramway Co, all operated horse buses.

In 1919 and 1920, the MR (NCC) introduced three AEC 23-seater petrol omnibuses to operate a Ballymena-Portglenone service in response to public demand. Later in 1920, a private operator, using smaller but faster Ford vehicles, went into competition with the buses on this route and as a result, in 1922, the NCC ended their service. In 1924 the LMS (NCC) converted bus No 3 into a railbus. It ran in this form for six months, operating a service between Coleraine and Portrush. After this experiment it was converted back to a road vehicle. The three AECs were sold in Belfast in 1925.

In 1924 the NCC bought a 14-seater Ford charabanc to replace the horse buses on the Shore Road route, which had been subject to competition from motor buses since 1923. This service was not a success and the vehicle was sold to McNeills of Larne in 1925. In Portrush, the NCC had been using horse drawn vehicles to operate from its Northern Counties Hotel. In 1925 these were replaced with a Star 14-seat saloon bus. This was the last service to be started by the NCC before private bus companies were acquired under the 1927 Northern Ireland Railways (Road Services) Act.

Great Southern and Western Railway Commercial Cars' charabanc in difficulties in Co Kerry.   Author's collection

GSWR Lancia Tetraiota coach.
Author's collection

The GSWR

After obtaining the legal powers to operate road services in 1903, in the same year the GSWR began a road freight service over a ten-mile route between Tuam and Dunmore. This was worked by two Thornycroft steam lorries, bought second-hand from Guinness's Brewery in Dublin. The service was not a success and was abandoned after a few months. A similar road freight service, introduced in the same year between Foynes and Newtown-Sandes in Co Limerick, was also a failure.

The company then turned its attention to the possibility of running charabancs. They made enquiries of Thornycroft, Wolseley, Milnes-Daimler and Straker-Squire to supply and operate vehicles "similar to the GWR". Wolseley informed them that they were unable to run a service on behalf of the GSWR and nothing more was done to follow up this idea.

These were the only initial direct ventures of the GSWR into the operation of road transport. Passenger services, which connected train services with the railway's hotels and which were advertised in the company's literature, were operated for them by contractors until after World War I.

In 1909 a number of 'motor cars' were hired from a firm in Macroom, on behalf of the GSWR, to operate tours in Co Kerry. In 1910 the company entered into an agreement with Tourist Development (Ireland) Ltd to operate charabanc tours in Kerry, on the railway's behalf. Tourist Development Ltd placed a number of

Commercial Cars charabancs into service. However, there weren't enough of these to carry the available traffic, so the GSWR agreed to buy six similar vehicles and Tourist Development would then buy them from the railway, on hire purchase terms.

In 1915 Tourist Development Ltd got into financial difficulties and sold the railway's six Commercial Cars charabancs to another company. This company fared no better financially so, in 1917, the GSWR repossessed the vehicles, although, as it turned out, none of the repossessed buses were of the original six bought by the railway company! After this, the GSWR stopped operating road services in Co Kerry until after World War I.

In 1924 the company began operating tours again in Co Kerry using four Lancia 14-seat coaches. In the same year it ordered two 18-seater Karrier charabancs for tourist work. These were not delivered until 1925, by which time the GSWR had been absorbed into the GSR. However, the Karriers were delivered in the GSWR livery with the company's crest on the sides.

The Commercial Cars charabancs were built on 22-seater WP1 or 14-seater WP3 chassis. They carried Bedfordshire or London registrations. Some vehicles, added to the fleet in 1911, carried Co Kerry numbers.

BCDR Dennis bus, with BCDR-built body, posed at Queen's Quay station, Belfast, with some important looking gentlemen on board.
Author's collection

Major road service operators

The BCDR

By 1922, the BCDR found that it was coming up against increasing road competition from small private operators, which were providing services that duplicated those of the railway, but at cheaper fares. The BCDR was in a particularly difficult position because its railway routes were closely paralleled by the main roads. In some cases, the road route between two towns was shorter than the equivalent rail route. Moreover, the bulk of the railway's traffic was passengers; there were no significant freight flows along the system. Another problem was that buses could operate into the centre of Belfast, whereas the trains terminated at Queen's Quay station. This was a good 20 minutes walk from the city centre, although there was a connecting Corporation tram service which operated right into the station complex.

In 1924, Matthew Morrow's Enterprise Bus Co introduced a service between Bangor and Belfast, using solid-tyred charabancs. There were four runs each day, three via Clandeboye and one via Crawfordsburn. The return fare of 1s 6d compared well with the third class rail return of 2s 4d. In 1925, encouraged by the success of Morrow's service, a rival service was introduced by the Bangor Queen Co.

A similar picture was emerging between Belfast and Ballynahinch, where the distance by road was six miles less than by rail. Two bus services were begun in 1922, charging cheaper fares than the railway. Although one of these withdrew from the market, three others entered in sequence, each new operator coming into the market as a predecessor withdrew, unable to make the service pay. As can be imagined from this, competition between road operators was intense.

Although, as outlined above, some of these road operators eventually were persuaded to act as feeder services to the railway, by the end of 1927, 27 regular rival bus services were competing in the area of County Down directly served by the BCDR.

This period saw a fares war waged, both between the various railway companies in the province and bus companies, as well as between the rival bus companies. In 1927, the BCDR submitted a document to the government detailing how this competition was adversely affecting them. In one paragraph it stated that:

> Great difficulty is being experienced in competing for traffic, in view of the fact that these motor bus owners have the facility of putting on services at a fare which could not under any circumstances be remunerative, and which must end in disaster to themselves.

Also in 1927, the BCDR began to operate new bus services of its own. The first of these, using two 32-seater Leyland Lion buses, was from Donaghadee station down the Ards peninsula to Ballywalter. In the same year, a 14-seat Dennis G and, in 1928, a further Leyland PLSC3 were purchased. In the winter of 1928 the BCDR began to operate the busy Belfast-Holywood road, using the Leyland PLSC3.

In 1927 the Railways Road Motor Act was passed and in 1929, under the powers granted by the Act, the BCDR began to buy out the opposition on the Holywood route. The first was H Russell of Holywood with five buses. Using these, in addition to its own vehicles, the railway company established a 15-minute service frequency on the Belfast-Holywood route. In June 1929 Gastons Motor Services, with its three AEC buses, and S Gillespie of Holywood, with two Reos, were acquired. With these, the BCDR was able to reduce the Holywood service to about a ten minute interval. In addition, the timings on the BCDR's original Newcastle-Kilkeel route were improved.

Most of the buses taken over from these operators were eventually replaced by additional Leylands and Dennises. The BCDR buses terminated on the Laganbank Road, close to the eventual site of the UTA's Oxford Street bus station. However, in spite of these moves, the BCDR found it difficult to recapture the traffic it had lost to the independent bus services. It did not have the financial resources to buy out all its competitors over its whole operating area, and on important routes like the Belfast-Newtownards run, the trains still faced intense competition from private road operators.

The BCDR was slow to recognise the threat and

BCDR Leyland PLSC1 used on the Newcastle-Kilkeel service.
Ribble Enthusiasts Club

potential of road transport. This lack of vision in its management, coupled with the disadvantages imposed by the geography of its operating area, made it especially vulnerable. Once the rot had set in, the BCDR never recovered and it was the first of the 'big three' railway companies in Northern Ireland to fall after World War II. Most of its network was closed down by the UTA in 1950 and the railway services replaced by the Authority's buses and lorries.

The BCDR acquired a number of buses from the four operators that it took over. From H Russell, of Holywood Co Down, it obtained four new Vulcans, between 1926 and 1929. The two 1926-27 buses were amongst the first Vulcans produced, 20-seaters of conventional layout. The later two BCDR Vulcan VWBLs, new in 1928 and 1929, were forward

BCDR Leyland LT1, new in 1930.
Ribble Enthusiasts Club

control 32 and 31-seater models. Vulcans were cheaper than the vehicles of their main competitors, such as Leyland and AEC, but they were not as satisfactory as the products of those companies.

From S Gillespie of Holywood the BCDR acquired two Reos, new in 1927; from J Gaston of Belfast three AECs, two new in 1926 to McClements of Belfast and the third bought by Gaston in 1929. Reos were American vehicles

which, in the 1920s, became the most successful US chassis to be sold in the UK. The company name came from the name of the founder, RE Olds, who had also founded the Olds Motor Works, later to become Oldsmobile, before being ousted in a boardroom coup in 1904. From this, he went on to found the Reo company which not only built chassis, but its own engines, and continued to sell vehicles in the UK until 1939.

Of the other vehicles that the BCDR acquired in the late 1920s and early 1930s, two Leyland PLSC1s were built in 1926; a Leyland PLSC2 and a Dennis 30 cwt in 1927; a Leyland PLSC3 in 1928; a Dennis E in 1929; a Leyland LT1 and an LT4 in 1930. Four Dennis Lancets were obtained, two built in 1932 and two more in 1934. Apart from one of the 1926-built Leyland PLSC1s, acquired in 1930 from Baxters of Airdrie in Scotland, the remainder of these buses were bought new.

Fourteen buses, built between 1926 and 1934, were transferred to the NIRTB in 1935. These comprised six Leylands, six Dennises and two Vulcans. All, except one Dennis 30 cwt 14-seater, had between 31 and 36 seats. Two Dennis Lancets were 36-seater coaches. All were withdrawn by the NIRTB by 1946. The smaller buses, nine 20-24 seaters, were disposed of, mainly to cross-channel operators.

Early vehicles carried the letters 'BCDR' in large letters along the side, with the company's coat of arms between the letters C and D. The early buses used on the Newcastle to Kilkeel run were painted in coaching stock crimson lake, lined out in gold. Later vehicles carried the company's name in full on the waist band below the window, with the coat of arms on the middle side panels. The later BCDR bus livery was light blue-green and white, with black mudguards and wheels.

The LMS (NCC)

The flood of vehicles which appeared on the roads after 1918 allowed the public to enjoy the benefits of cheap, door-to-door transport and this began to hit railway traffic increasingly hard. By 1923, when the LMS took over the NCC, competition was especially severe around Belfast, Ballymena and Larne, even though the railway fought back with more competitive fares and improved timetables.

In 1925, it was estimated by the NCC that about 47 buses were operating between Belfast and Larne, on a road closely parallel to the rail route. James Pepper, the NCC's general manager, requested that his trains be allowed to operate from the cross-channel steamer terminal at Donegall Quay, over the harbour tramway lines, because private buses meeting the boats were taking so much of the traffic. It is not thought that this idea proceeded beyond the stage of running a train to test the route and it is probable that bogie passenger coaches were not suitable for operating over the tight curves of the harbour lines.

In 1927 HMS Catherwood, one of the largest private bus operators, brought out a summer timetable giving an hourly service from Belfast to Portrush and four buses per day from Belfast to Derry, both prime NCC routes. Although the speed and standard of comfort of the buses on the roads of the time could not have matched that of the NCC's modern carriages, the cheap fares inevitably persuaded a large number of passengers to use the buses.

In 1926 the new Northern Ireland government passed an Act licensing road vehicles and their crews. This had the effect of encouraging small operators to create larger combines, because no licences were to be granted after 1928 to new operators without government approval. This had the effect of protecting existing operators.

In 1927, a new combine of the principal bus operators was formed – the Belfast Omnibus Company, or BOC – which by 1929 owned over 200 buses. In that year, Catherwood and the BOC issued tourist programmes. Catherwood used the slogan, "First class travel for less than third class fares", an obvious dig at the railways.

In the same year, the Railways (Road Vehicles) Act was passed. This allowed railways, wholly situated in Northern Ireland, to own and operate road buses. Although the NCC had such powers already under the 1899 Act and, as has been noted above, had operated buses on a small scale during its BNCR and MR days, it was only now that the thought of buying out the road competition really took hold at York Road.

Later in 1927 McNeills of Larne, who were

Top: *LMS (NCC) Leyland Lion at Culmore on the tram replacement service to Portstewart. This was an ex-BOC vehicle.*   CP Friel collection

Centre: *LMS (NCC) Smithfield bus station, Belfast, in the 1930s.*
Author's collection

Bottom: *LMS (NCC) Leyland Tiger No 97.*
Author's collection

contracted to the NCC for coast road work, were approached by the BOC with a view to selling out to them. Pepper recommended to the NCC board that a counter-offer be made to McNeill to buy his operation. In the event, this was not followed through, as McNeill did not want to sell.

By 1929, Pepper had come to the view that buying out the competition would not only eliminate it, but would allow economies to be made, by replacing uneconomic rail services with the NCC's own buses. Later in the year, he reported to the NCC board that he, along with the GNR and BCDR, had made a joint approach to the BOC with a view to buying it out. He indicated that, if this was successful, the Ballyclare, Draperstown, Dungiven, and Portrush branches could be closed and replaced by buses and that the Cookstown and Derry Central line services could be reduced. His board gave him the authority to proceed and the NCC began to buy buses and routes from private operators.

The first service the NCC bought was that between Belfast and Kilroot, operated by H Martin of Kilroot. In the same year they bought a Belfast to Ballyclare service of eight buses. By the end of the year, 22 buses had been acquired. The NCC also allowed bus return tickets to be used on trains.

In 1930 the NCC's most important acquisition was the BOC's routes in their area, along with 59 buses. Twenty-five buses were also bought from three smaller operators, including 12 vehicles from McNeills of Larne. By March 1930, the NCC owned 118 buses and ended up the second largest bus operator in Northern Ireland, the BOC beating it by just a few vehicles. By the end of this process of buying out the competition, the NCC had taken over 18 private operators.

As a result of the way it acquired its bus fleet, the NCC ended up with buses of a great variety of makes, ages and conditions. It set about reorganising and integrating its road services and ordered 23 new buses to replace life-expired machines. York Road works was extended and equipped for carrying out heavy repairs. The 19 provincial depots it had acquired were reduced in number to 13.

The NCC originally used part of the BOC's Belfast yard in North Street. However, in 1930, they bought the nearby premises of the New Smithfield Weaving Co, which was converted into Smithfield bus station. From 1931 this depot was shared with the GNR, which used it as its Belfast terminus. Smithfield depot, with its original LMS (NCC) bus shed and unusual single span roof structure, remained in use until the late 1970s. It was closed only after being badly damaged in a terrorist attack.

Bus services were re-organised to complement rail services and this allowed the NCC to reduce train mileage and close branch lines. Most of the NCC narrow gauge system was closed to passenger traffic between 1930 and 1933, and the standard gauge Ballyclare branch was closed to passenger traffic in 1938.

During the following years, the NCC bought a considerable number of new buses and standardised eventually on a diesel-engined bus fleet. The first bus stop posts, made of the ubiquitous LMS concrete, were erected on the Shore Road in Belfast in 1931 and a few of these were still around 30 years later. In the same year new garages were opened in Derry, Cookstown, Magherafelt and Whitehead, the latter an important depot for regular and tourist services on Islandmagee, a peninsula close to Whitehead.

In 1932, for the joint NCC/GNR tours, each company purchased two Leyland six-cylinder 'sun saloons', with armchair seats for 24 passengers. Belfast city tours were developed, as well as private hire work, and all-Ireland tours were run jointly with the GNR.

Garages were built in Kilrea, Ballymena and Ballyclare in 1932, and bus services provided between Ballymena and Coleraine railway stations and their respective town centres. Most NCC bus depots were built near railway stations, on land taken over from railway use. Four further small private operators were also acquired. In 1934, Limavady engine shed was converted into a garage and a new large garage was completed in Coleraine in 1935, the last major improvement to its bus provision carried out by the NCC.

NCC buses were painted in LMS crimson lake and white and lettered 'Northern Counties'. This was similar to the original BOC colour scheme. When the NCC acquired the right to use this livery, the BOC switched to two-tone green, a colour combination that was to play an important part in later road/rail

LMSR Fleet number	Reg. Co	NCC Fleet number	Make and Type	Chassis number	Body	Seating
11F	CH7800	?	Leyland PLSC3	47395	LMS	B32R
19F	CH7910	42 (railbus)	Leyland PLSC3	47403	LMS	B32R
22F	CH7913	?	Leyland PLSC3	47406	LMS	B32R
202F	CH7917	?	Albion PM28	7029L	LMS	B32R
207F	CH7922	?	Albion PM28	7032K	LMS	B32R

Above: *Details of LMS buses transferred to the NCC in 1935.*

relationships. However, BOC crew uniforms remained brown. NCC bus crews were also provided with a standard uniform.

The buses had white roofs, window surrounds and waistband. The lower halves of the vehicles were red with black mudguards and wheels. The words 'Northern Counties' (in capitals) were applied along the waistband, below the windows. Fleet numbers were carried on the front and back only, at waistband level. The LMS coat of arms, displayed on railway vehicles, was not carried by the buses.

In 1935 five buses were transferred from the LMS bus fleet in England (see table above). They comprised three Leyland PLSC3 Lion and two Albion PM28s, which had been used on the Rochdale-Halifax route. Two of the Leyland Lions were converted into railbuses for use on the Coleraine-Portrush branch. The third Lion was converted into a parcels van, while the Albions were converted into goods lorries.

One of the Leylands is pictured on page 106 in its railbus form, with its English registration plate still in place but with the fleet number 42. This was not its English fleet number, but seems to be a fleet number in the NCC bus fleet list. In the fleet list of NCC buses transferred to the NIRTB, there is a gap in the early 1940s.

In spite of the NCC's successes, the general road transport situation, from the railways' point of view, was not good. Although a proper system of licensing both vehicles and operators had allowed the railways to virtually eradicate private bus opposition where it was hurting them, the same was not true for freight. Here the position was similar to that which had prevailed in the bus industry before 1926.

Malcolm Speir, with the support of the GNR and

the BCDR wanted legislation passed to allow the railways to become 'Transport Companies' with powers to buy out private freight operators and set up fully integrated transport operations in their areas. However, after meeting deputations from the railways' top managements, the government decided, in 1934, to appoint Sir Felix Pole, the ex-general manager of the Great Western Railway in England, to consider the whole transport position in Northern Ireland.

As we have seen (page 17), Pole proposed the establishment of a 'Northern Ireland Road Transport Board' to integrate all road freight and passenger operations, including railway bus and freight operations. The board would co-operate with the railways to provide an integrated service provision, and to pool receipts with them. The railways, by and large, agreed with this scheme as a way out of the problem.

The NIRTB was established in 1935 and, in that year, the NCC handed over to it 131 buses, 56 lorries, their operating and maintenance staffs, Smithfield bus station and other facilities. The NCC became once again purely a railway system and soon found that, rather than co-operating with the railways, the new NIRTB was competing with them. The road/rail problem had not been solved and the NCC, as with other railways in Northern Ireland, was now worse off than before.

The GSR

Soon after the Irish Free State was established in 1922, private bus operators began to spring up all over the country. As in Northern Ireland, competition between operators and between them and the railways was intense, and the railways began to suffer

*Great Southern Railways Leyland TS2
No 406, built 1929 for the IOC as
their No 30.* Author's collection

significant damage from these unregulated road operators.

On 1 January 1925 the GSR was formed, by the grouping of all the railways operating totally within the Irish Free State. The GSR ran bus services under its own name with a fleet of charabancs which operated on the following routes:

Clifden-Mallaranny
Sligo-Bundoran-Ballyshannon
Kenmare-Parknasilla
Macroom-Glengarriff
Killarney-Kenmare-Glengarriff-Bantry

These, however, were only tourist services and operated in conjunction with the trains during the summer season only. The buses they took over comprised four 14-seater Lancia coaches used on the summer tours in Kerry, and three Karrier 14-seaters used on the summer tours in Connemara.

The GSR board considered whether winter work could be found for these vehicles. No decision on this was taken immediately, but it was later decided that co-operation with existing coach operators was necessary and that services be bought in with hired buses. The GSR did toy with the idea of buying some Ford buses of its own, but decided against it. By 1926 a number of routes were being operated on a contract basis and some firms were taken over, but at this stage, in no case, did the GSR actually supply or operate its own buses.

In 1926 the Irish Omnibus Company (IOC) came into existence. It originated with the Clondalkin Motor Omnibus Company, which had been set up in 1919 to operate between the Inchicore tram terminus and Clondalkin. The Clondalkin company was taken over in 1926 by FT Wood of Altrincham, Cheshire, and in December of that year the IOC was incorporated to take over and expand the Clondalkin operation. Within a few months, the IOC was operating 16 buses over ten routes.

In 1927 the Railways (Road Motors) Act was passed. This empowered the railways to operate buses, subject to ministerial approval of proposed routes and charges. As a result of this, the GSR and the IOC reached an agreement under which the IOC became the agent of the GSR and took over the GSR's road licences. The IOC began to operate throughout the Irish Free State.

In July 1929 the GSR acquired a controlling interest in the IOC. Initially the railway continued to operate its summer tours and, in the summer of 1929, ordered three Karrier 18-seat coaches in addition to those it already had in service. However, by the end of 1929, it was agreed that the IOC would also take over the GSR's charabanc and coach services.

In September 1929 inter-availability of tickets between road and rail services was introduced. Broadstone station, the erstwhile headquarters of the Midland Great Western Railway, was partly converted into a bus garage. Three tracks on the arrival side were filled in and only one line was left for both arrivals and departures and a line for the storage of carriages.

However, the GSR was still not satisfied with the position it was in vis-à-vis road passenger and freight competition and, in 1931, the chairman of the GSR called for fresh legislation "regulating transport and removing the disabilities at present imposed on railways".

By that time, the number of licensed road vehicles had increased by 66% to 49,000, compared to 29,000 in 1925 when the GSR was set up. Road transport was virtually unregulated and the situation was getting out of hand.

In 1932 the government introduced a Road

GSR Leyland TS7 streamlined coach No 216, built 1936, posed outside Broadstone station. Author's collection

Transport Act which stated that all scheduled passenger services must be operated under licences issued by the state. Timetables and charges would also have to be published. The Act did not proscribe independent operators; they could continue and new ones could be set up, provided they met the requirements of the Act. In 1933 a further Road Transport Act, modelled on that in the UK, was passed. Under this Act railways were given the power to compulsorily purchase their road competitors, both passenger and freight.

As a result of these Acts, four operators – the GSR, the GNR, the LLSR and the Dublin United Tramway Co – ended up controlling most of the bus services within the Irish Free State. About 30 private operators were left, but between them they owned only about 100 buses, and none was left operating a trunk route. Only the Saint Kevin's Bus Service (which still exists) was left operating a stage carriage service into Dublin. The IOC, and its successor the GSR, consolidated and expanded their services and introduced coach tours and excursions.

In 1934 the GSR absorbed the IOC, which was then wound up. It

GSR Leyland TS6 No 750, built in CIÉ colours. Photobus

became the Omnibus Department of the GSR

During World War II, shortages of petrol forced the GSR to reduce bus services. This created a surplus of vehicles and allowed the GSR to send 31 buses to Northern Ireland, to replace NIRTB vehicles requisitioned by the military. GSR buses which were sent north were away for up to two years and were placed in a special NIRTB number series.

The war also resulted in a shortage of spare parts, which by 1944 required the GSR and DUTC to co-operate for their provision. The DUTC took 12 single-deckers out of service and the GSR 51 single-deckers. Most of these were totally cannibalised during the period before the formation of CIÉ. The bodies of four of nine cannibalised Leyland Lions

over by CIÉ. Ten Titans were built in 1940 and in 1942 three Tigers were built for the GSR by the DUTC.

The original fleet numbering system of the IOC is believed to have been in a straight series below 100. By about 1929 they renumbered the fleet according to type, each type being allocated numbers as follows:

were later sold to the LLSR.

By the time of its absorption into CIÉ, in January 1945, the GSR owned 254 operational buses and worked all the main provincial services south of a line between Dublin and Sligo, those north of this line being operated by the GNR. This was as the result of agreements entered into in 1932 between the IOC and the GNR, and between the GNR and the DUTC. The GSR also operated the Cork, Waterford and Limerick city Services.

A great variety of makes of vehicles were acquired from private operators over the years but, from 1929, only Leylands were bought new. IOC bus bodies were built by contractors or, until 1934, by the GSR at Inchicore. In that year bus building was transferred to Broadstone, where it remained until 1940.

Originally, IOC/GSR buses were petrol-engined, but by 1940 they had switched over to diesels. The last new petrol-engined Leyland Lions and Leyland Tigers were built in Broadstone in 1940, and were the last buses to be built there. By that year, the GSR had standardised on diesel-engined Leyland Titan double-deckers and Leyland Tiger single-deckers with metal-framed bodywork. However, the bulk of the GSR fleet remained petrol-engined until take-

Below 100	Various acquired vehicles and new Leyland Lionesses 91-98
100-199	ADC and AEC
200-299	Albion
300-399	Leyland Lion PLSC
400-499	Leyland Tiger, Vulcan and AEC Regal
500-599	Leyland Badger
600-699	Thornycroft
700-799	Leyland Lion LT
800 up	Leyland Titan

The GSR continued with the IOC's system, amending it as follows:

100 up	reissued for Leyland Tiger
550-579	acquired Bedford
580-589	acquired Commer
590-599	acquired Gilford
650-699	Leyland Cub
900 up	Leyland Lion LT

When CIÉ was set up in 1945, the fleet taken over comprised 34 double-deckers (Leyland Titans) and 220 single-deckers, made up of three AEC Regals and 217 Leylands (Cubs, Lions and Tigers). Only 14 of the double-deckers and 11 Leyland Tigers had diesel engines.

The livery of IOC buses was signal red lower panels, ivory upper panels and black lining for service buses; ivory bodies and roofs with black lining for coaches. The GSR retained this livery. Most vehicles in IOC days carried a gold transfer on the side, consisting of a large oval surrounding the letters IOC (a similar device was used by the Belfast Omnibus Company). When the GSR took over, their initials replaced those of their predecessor in this oval, sometimes using the Irish equivalent of the initials GSR, with 'MI an D' in Gaelic script. This stood for 'Mór Iarnród an Deisceart' or literally 'Large Railway of the South'. However, several buses wore the GSR's heraldic coat of arms, the same as carried by railway vehicles. Detachable black-on-white bi-lingual destination boards were carried above the cantrail.

The CDRJC

The CDR had the advantage over most other railway companies in Ireland, that during the period up until the late 1930s, the roads in Co Donegal were so bad that competition from road transport was not really a serious proposition. In south and east Donegal, where the CDR mainly operated, early conditions favoured the railways which linked the various settlements together more efficiently, in many cases, than did the roads. The CDR, unlike its north Donegal counterpart, the Londonderry and Lough Swilly Railway, had its stations well sited in relation to the towns and villages it served.

Nevertheless, the CDR management did see the advantage of bus feeder services to the railway. In 1929, they applied to the Ministry for Industry and Commerce in Dublin for permission to operate three bus routes – Glenties to Ardara, Glenties to Portnoo and Killybegs to Glencolumbkille. This approval was given.

The CDR then bought four second-hand 36hp petrol-engined Reo buses in 1930 from the GNR to operate the routes. These four buses lasted until 1933, by which time the poor roads had totally worn them out. As noted in the chapter on railbuses, the two best were converted into railbuses Nos 9 and 10 in the CDR railcar fleet. This means, of course, that the first American internal combustion-engined railway vehicles in Ireland, were not CIÉ's GM locomotives

of 1961, but the CDR Reos!

After this brief excursion into bus operations had ended in 1933, bus services in County Donegal were provided solely by the Great Northern Railway. As it turned out, the GNR Donegal bus services were, for the most part, loss-makers. In 1954 the GNRB employed a firm of consultants to assist it in carrying out a strategic review of its road passenger operations. In this review, it was recommended that the loss making Donegal services be discontinued. The road services of the GNRB were absorbed into CIÉ in 1958 before this recommendation could be carried out.

In 1947, the CDR management decided to get involved again in the provision of bus services as a result of the closure of the Glenties branch. Four routes were operated – Stranorlar-Glenties-Portnoo, Glenties to Dungloe, Ballybofey to Letterkenny and Malinmore to Killybegs. These routes were operated, not directly by the CDR, but by the GNR, on the CDR's behalf. At Dungloe they met the bus services of the other major transport operator in the county, the LLSR, which in 1953 would abandon the use of railways in favour of a total road provision.

The final phase of CDR involvement in bus transport began in 1960. By 1957, it was clear that the railway could not be run profitably and the decision to terminate all rail services was taken. The last trains ran on 31 December 1959 and substitute bus services took over the next day.

From 1 January 1960, all CDR passenger services were operated by six P-class Leyland 39-seater single-decker buses, hired from CIÉ and painted in the CDR's geranium and cream livery. Four or five additional P-class vehicles were hired to cope with the increased summer traffic. As far as possible, the bus routes paralleled the old railway routes, ran to the same timetables and in some cases used the railway stations as the boarding and setting down points in the towns and villages they served.

The hired CIÉ P-class buses lasted until 1965. A decision had been taken by the CDR management the previous year to operate their own vehicles rather than hired ones. Six Saro-bodied 44-seater Leyland Tiger Cubs were acquired in 1965 from the East Midland Road Car Co in England. They were overhauled by the CDR in their Stranorlar workshops, and were painted in the CDR red and

CDR Leyland OPS3 (CIÉ P-class No 31 on hire) at Strabane station.
WH Montgomery

cream livery, some by the CDR and some by O'Doherty's coach works in Strabane.

One odd feature about these buses was that they never carried fleet numbers. They were identified throughout their lives by the English registration numbers which they retained while in CDR ownership. They gave good service and formed the core of the passenger carriage provision. In the winter, they were supplemented by a hired CIÉ E-class single-decker and in the summer by six hired CIÉ vehicles, a mixture of E and P-classes. In 1966 the Tiger Cubs were converted to one-man-operation.

All the Tiger Cubs were withdrawn by 1970, after which all services were operated by a combination of CIÉ E-class Leyland Leopard buses, and school services Bedfords.

In partnership with the newly formed Ulsterbus, the CDR introduced two express coach services. The first was the Belfast to Killybegs express, begun in the summer of 1967. The Killybegs to Strabane section was operated by the CDR with their own vehicle. This service was time-tabled to connect with CIÉ express bus services which operated from Dublin and Galway. The second express service, introduced in 1971, was the Letterkenny to Aldergrove (Belfast Airport) service, operated as a joint service with Ulsterbus, just before the CDR was absorbed into CIÉ.

In addition to normal stage carriage work and express services, the CDR also operated school bus services in their area on behalf of the government of the Republic. A government grant was provided for such work.

On 12 July 1971, the CDR was officially absorbed into CIÉ. As the last Tiger Cub had gone out of service in 1970, the last vehicles to have been operated by the County Donegal Railway Company were the E-class Leyland Leopard buses.

It is believed by some observers that the CDR missed the opportunity to convert fully to road operations in 1947, when they used the GNR to provide the replacement services for the Glenties branch. Bus services seem to have been simply 'bolted on' to the rail provision, up to the point when it was clear that the railways could not survive at all. Even after 1960, most of the CDR's bus operations were provided using buses hired from CIÉ. A more strategic view might have allowed the CDR to convert gradually to total bus operations over a longer time scale using their own vehicles, much as the Lough Swilly did, and possibly to have survived longer as a separate entity as a result.

The bus livery used by the CDR was the same as that for the railcars and passenger carriages; geranium red below the waist band and cream above, with the company's coat of arms on the side panels.

The LLSR

After World War I, the Lough Swilly Railway, like many other lines, found itself saddled with high operating costs, as a result of the high pay rates introduced when the railway was under government control.

By the mid 1920s, now with the additional problems posed by the existence of an international border just a few miles beyond its headquarters in Derry, the company was in a deteriorating financial position. However, the terrible condition of the roads in County Donegal kept road transport competition at bay for longer than in the more prosperous east of Ireland. In spite of this, from 1924 onwards, state subsidies from the governments of both Northern Ireland and the Irish Free State were necessary to keep the lines open.

Right: LLSR 18-seater Morris No 40, acquired from the Royal Artillery at Dunree Fort in 1938.
Author's collection

Below: LLSR Leyland PD1A as new, posed near Buncrana railway station.
Ribble enthusiasts club

In 1927, the Lough Swilly approached the County Donegal Railway with a view to the latter taking over their system but, after negotiations, the CDRJC decided against this move.

Although the neighbouring CDR was using railcars extensively by the early 1930s, the Lough Swilly waited until 1931 before even considering their use. They examined the CDR's railcar operations and even obtained a quotation from Armstrong-Whitworth for two petrol electric rail-motors, on which they took no action, probably because by this stage they realised that the future for them lay in road transport.

The turning point came in 1931 when the company came up with a plan which was approved by both the Irish Free State and Northern Ireland governments. Under this plan, the Swilly would acquire competing, privately owned road services. The idea was that this would allow the company to eventually close the railway completely and operate totally with road vehicles.

This scheme, as it was gradually implemented, quickly led to an improvement of the railway's finances. The railway strike in 1933 accelerated the transfer process and, by 1935, the company had acquired lorries and buses and established an organisational structure, designed to eventually replace that of the railway. Legislation was passed which allowed the closure of any section of the railway, once the company had made adequate provision for the carriage of goods and passengers by road.

During World War II, the process of moving to road transport was put into reverse to some extent. Rail services were partially restored on lines from which they had been withdrawn and loadings went up, particularly between Derry and the nearby cross-border Donegal seaside resort of Buncrana, which of course was in the neutral Irish Free State. With the end of hostilities, and the resulting easing of supplies of oil, the run down of the railways recommenced and in August 1953 the last train ran between Buncrana and Derry.

With regard to the details of bus operations, the Swilly's first bus was acquired in 1929 when it purchased the service of a Mr Barr of Buncrana. However, as noted above, it started buying private operators in earnest in 1930. By the end of 1930, it owned 24 vehicles and, by the end of 1931, 37 buses. As the result of scrapping, the number of buses dropped to 32 during the war but, by 1953, when train services ceased, the number of buses had increased to 52.

Financial constraints have heavily influenced the Swilly's bus purchasing policy since the beginning and has resulted in the company having to rely on second-hand buses, from both Irish and British operators, for much of its fleet. This, however, has made it a fascinating study for transport enthusiasts down to the present day.

The Swilly is in the anomalous position of being a company registered in Northern Ireland, but with its operating area almost totally within the Irish Republic, in Donegal. This prevented the original road operations being absorbed into the NIRTB in 1935, or the railway and road operations being absorbed into the UTA in 1948.

For the same reason, the LLSR and the neighbouring CDR, which also operated mainly in the Irish Republic, were not absorbed by the Great Southern Railways in 1925, with the rest of the Irish Free State's totally indigenous railway companies. The same situation applied in 1945 when CIÉ was established in the Republic of Ireland to operate that state's rail and bus services. Therefore, the continued independent existence of the Londonderry and Lough Swilly Railway Co is one of the consequencess of Irish politics and history.

Old railway stations were used as bus stations and the Pennyburn and Letterkenny railway engineering facilities were gradually adapted for servicing the bus fleet. Very little change took place to these facilities until recently. Up until the 1970s, buses were still being repaired in the old Pennyburn railway workshops in Derry, sitting over pits still flanked by three foot gauge railway track. In Letterkenny, buses were based in premises consisting of the remnants of the railway stations and railway yards of both the Lough Swilly and the CDR right next door. Bus conductors with the Lough Swilly continued to be known as 'guards', railway-fashion.

Because the Swilly operates on both sides of the Irish border, crews have to hold Public Service Vehicle licences from both states. Buses also have to satisfy the Public Service Vehicle regulations of both states and vehicle insurances have to be valid in both countries.

Because we are primarily interested in railway companies which operated road vehicles, and because the company is still in existence, it is intended to draw a line under the Londonderry and Lough Swilly Railway Co's bus operations in 1953, the year the Swilly ceased to operate trains.

The LLSR bus fleet, since the beginning, has been a very miscellaneous group of vehicles. Some buses were bought new, but many more were acquired second-hand, initially from operators taken over in Donegal and in later years bought from companies all over the British Isles.

As already noted, the first vehicles that the LLSR acquired were from Barrs of Buncrana in 1929. They were three Graham-Dodge buses and one 34-seater Leyland LT. In 1930 it bought several new buses, four more Leyland LTs and one Leyland TS3. No further new vehicles were bought until 1933, though during the intervening three years the LLSR acquired 34 second-hand buses from a variety of private firms it had taken over. These buses included Gilfords, Reos, Guys, Chevrolets, another Graham-Dodge, an Albion PK26 and, that rarity, a Morris bus.

In 1932, the LLSR acquired their first batch of second-hand buses from mainland British operators. The first of these were seven almost new 32-seater Vulcans from Steele and Ferguson of Glasgow. In 1933, two new Leyland TS3s were bought and, in 1934, two Leyland PLSC3s second-hand from the Belfast Omnibus Company.

Two new Leyland LT7s were bought in 1936 and an 18-seater Morris was acquired in 1938 from the departing Royal Artillery Regiment. This regiment had been based at Dunree Fort at the mouth of Lough Swilly. Two ex-Dublin United Tramways Company vehicles, an AEC Regal and Leyland SKP3, were acquired in 1939.

During the war, the Lough Swilly managed to acquire seven buses. The first six – two Albion PK115s, an Albion PW65, an Albion CX9, a Bedford OWB, a Leyland LT5A – all came from The Major bus service in Belfast, which had just been absorbed into the Belfast Corporation's undertaking. The seventh, a Leyland LT4, came from the NIRTB. This

LLSR Saro-bodied Royal Tiger No 71, delivered in 1951.

Ribble Enthusiasts Club

bus had been new to the LMS (NCC) in 1932.

Three new Bedford OB buses were acquired immediately after the war. From this time the company began to standardise on the Leyland heavyweight chassis for both bus and freight vehicles, in both its new and second-hand purchases. This was because of the poor condition of Donegal's roads, as well as the possibility of standardisation and inter-changablility of parts between passenger and freight vehicles.

In spite of this general policy of standardisation, in 1946 the company took delivery of eight AEC Regals fitted with second-hand bodies, seating from 30 to 34 passengers. In 1948 four more AEC Regal IIIs were purchased.

Double-deckers made their first appearance in 1947, in the form of two Leyland Titan PD1As. These were fitted with Alexander bodies to Leyland design. They were shipped, in knocked-down form, to Derry, where they were assembled and fitted. These buses were used on the heavily trafficked Derry-Buncrana route and on schools services.

In 1949 four Leyland Tiger PS2/1 single-deckers went into service. These had been built for the Swilly by the UTA at their Dunmurry workshops, just outside Belfast. In the same year, the company's next double-deckers appeared – Leyland PD2/1s with Leyland-designed and built bodywork.

In 1950, a second batch of new UTA-built Leyland PS2/1s were bought, plus ten second-hand buses, nine of which were ex-UTA vehicles. These buses were a mixture of Leyland LT7s with Leyland bodies, Leyland Tiger TS6s with NIRTB bodywork, four Bedford OWBs and a Leyland PD2/1 double-decker demonstrator. The demonstrator had also spent some time in service with the Belfast Corporation Transport Department.

One of this batch, No 13, was a Leyland TS6, which had been NIRTB/UTA No M402. It had started life as an LMS (NCC) bus, No 14, with a 32-seat

Weymann body. This body had been removed by the NIRTB in 1944 and replaced with a standard NIRTB B34R body. The LLSR disposed of this bus in 1958, but it survived in private hands and has now been preserved, fully restored into NIRTB livery.

In 1951 the first underfloor-engined buses in Ireland were introduced by the Swilly. These were four Leyland Royal Tiger PSU1/9 single-deckers. They had been ordered originally for use in Argentina, but bought by the Lough Swilly when this order was cancelled. The buses were fitted with Saunders-Roe (Saro) bodywork and were so good that they lasted in service until 1979.

From 1953 the Londonderry and Lough Swilly Railway Company became a purely road operator. Four more Saro-bodied Royal Tigers were added in 1953 and one of this batch, fleet No 75, has survived into preservation.

The wisdom of the LLSR policy of replacing rail with road operations so early, cannot today be doubted. The company has survived under its original 1853 Act of Parliament, the last of the cross-border oddities, still operating in the area the original railway was built to serve. Its policy of operating on a shoe-string, using buses acquired second-hand from Ulsterbus, CIÉ and British companies, particularly Ribble, has continued right down to the present.

The Londonderry and Lough Swilly Railway Co's bus fleet carried two liveries during the period covered here. For many years it was ivory and black, similar to the IOC/GSR coach colours but, in the late 1940s, it was changed to a khaki and green livery, with the words 'Lough Swilly' carried in green script along the bus sides. This livery lasted until the mid-1960s. Various other liveries have been carried since then.

The GNR

The road passenger operations of both the GNR and GNRB will be covered in this section, as the evolution of the GNR road provision was not seriously affected by the change in the structure of ownership of the company in 1953.

As with the other railway companies discussed above, the impact of road transport on the GNR began after World War I, with the influx of surplus army vehicles, the new products of an industry which had been geared up for war production. As elsewhere, the railway, highly regulated by government legislation, was faced with a totally unregulated road transport industry, able to offer flexible and cheap transport which the railway could not match.

The GNR traffic between Belfast and Dublin, and the heavy suburban traffic of both cities, was not seriously affected at first. Early road transport was generally not capable of offering a reasonable alternative to the trains. But the company did lose traffic on short country journeys, where access to farms and small villages was easier and better by bus and car.

The effects of competition from road transport began to make themselves felt on the company's profit and loss account. A growing surplus up until 1920 was replaced by a declining one during the 1920s and, by 1933, the railway's finances had moved into loss, a situation which did not change until World War II.

By 1927, the governments of both Northern Ireland and the Irish Free State had decided to pass legislation to allow railway companies to run their own road services. Once these Acts were passed, the GNR began the expensive business of purchasing its road competitors' operations and planning, in their place, a road service integrated with its railway provision.

As with the other railway companies in this position, the GNR acquired a motley collection of vehicles, many of which had to be put into good order. A policy of gradual standardisation was adopted. In 1934, prior to the establishment of the NIRTB, when much of its fleet was handed over to

the new Board, the GNR owned 170 buses. This was the largest number it ever had. During this period, the Great Northern built up a bus network which was especially strong on providing bus feeder links to its rail services, but also included trunk routes, tours and private hire work.

The Great Northern Railway's bus services commenced on 28 January 1929 with a town service in Drogheda. Powers were granted to the GNR, under the Railways (Road Motors) Act 1927, to operate a large number of services in the Irish Free State. The first long-distance route to be operated was in February 1929, from Dublin to Drogheda. This route had been operated by the Louth and Meath Omnibus Co, which the GNR had taken over. The Great Northern continued to extend its sphere of influence by taking over small operators on both sides of the border.

The first route it operated in Northern Ireland was the Belfast to Lurgan route, acquired with the take over of JR Irwin in March 1929. It also took over Irwin's routes to Banbridge, Portadown, Newry and Rathfriland. From Irwin it obtained a number of new Dennises. For these routes, and the company's own new services in the Free State, 20 AEC 32-seater buses were ordered. Five of these were used to replace Irwin's older buses operating between Belfast and Lurgan.

In 1929, the GNR also opened a service from Dublin to Malahide by two routes, one by Coolock and the other by Raheny, Baldoyle and Portmarnock. This was the first service to the latter two places. From July, they began to operate a service between Dublin and Skerries, this in addition to the service between Skerries station and town. In the same year, the GNR also took over seven buses owned by Parkes of Monaghan, operating between Monaghan and Dundalk.

In the early 1930s the Great Northern reached two important agreements. The first was concluded in 1930 with the Dublin United Tramway Company and gave the GNR the suburban bus services in the north of Dublin. The second, in 1932, was with the Irish Omnibus Company, which allowed the GNR to consolidate its bus network north of a line between Dublin and Sligo. Another important addition to the network, in 1933, was that of the long distance services of the Belfast-based HMS Catherwood,

GNR Albion PX65 No 246.

which were acquired compulsorily. This included Catherwood's Belfast to Dublin service.

From July 1930, the GNR introduced the inter-availability of road and rail tickets. In Northern Ireland, return tickets were made inter-available between bus and train except on the Belfast-Lisburn route, where special cheap fares were introduced. Where appropriate, this inter-availability also included the services of the Great Southern Railways and the Irish Omnibus Co, which was controlled by the GSR. After the foundation of the NIRTB, connections with its services were included.

In all, the Great Northern absorbed 31 independent companies between 1929 and 1945, 25 of them between 1929 and 1932. Eighteen of the total were in the Irish Free State and the other 13 in Northern Ireland.

In the Irish Free State, up to 1932, anyone could set up and run a road passenger service. However, in 1932 the new Road Transport Act, which repealed the 1927 Act, set up a licensing system which reduced the number of new entrants, improved operating standards and gave the railways extra powers to acquire road operators. The Act was intended to reduce wasteful duplication and uneconomic competition, which it did successfully. It allowed the GNR to acquire additional road freight competitors and to further integrate its road and rail operations.

However, the situation in Northern Ireland was different. Relatively unbridled road competition was allowed to continue so that, by 1934, the very survival of railways in the province was in question. As we have seen, the Northern Ireland government's solution to this problem was to establish the Northern Ireland Road Transport Board in 1935. The board was charged with the integration of the road transport provision in the province and the co-ordination of this provision with that of the railways. The GNR had to hand over 50 buses to the NIRTB and could no longer provide road freight or bus services, other than connecting services, north of the border.

In return for the road vehicles it had transferred to the new board, the GNR and the other Northern Ireland railway companies had been promised proper collaboration with the NIRTB. In the event this did not happen, the NIRTB simply becoming a more

*Right: GNR Dennis Lancet I No 235,
one of four purchased in 1932.*
                    Author's collection

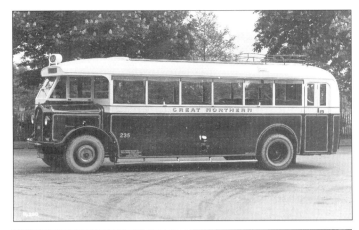

Below: GNR AEC Regent No 440.
                            Duffners

efficient competitor than the previous private operators had been. Like the other Northern Ireland railway companies, the GNR suffered as a result of this policy.

In the Irish Free State (the Republic of Ireland from 1949), the GNR road fleet expanded again to a maximum of 170 buses. In 1937 the first two double-deckers were acquired. By 1955 the fleet stood at 34 double-deck buses and 123 single-deckers.

In 1954, just after its formation, the Great Northern Railway Board employed a firm of consultants, Urwick, Orr and Partners, to help the board carry out a major strategic review of the road passenger services and their place within the total passenger transport provision of the company. This review was intended to form part of the basis of a ten-year corporate plan for the company.

The consultants' submission made some interesting observations and recommendations which, had they been implemented, would have dramatically changed the nature of the GNR itself and the passenger services it provided. The company would have ceased to be primarily a railway company and would have become more like the UTA, the structure of which probably influenced the consultants' thinking. It would have become, in fact, a transport company with road provision the dominant element and railway services playing a secondary role.

The report noted that the peak profitability of the road services was attained in the years 1947-49. Subsequently, the introduction of CIÉ conditions of service in 1951, and a strike in 1952, converted a considerable profit into a small loss. In 1953 and 1954 the service moved back into profit, though only at about 25% of the 1947-49 levels.

It was also noted that the policy of the GNR had been to operate its road provision in support of, and/or in protection of, its rail activities, with the intention of providing an integrated transport service. Significantly, it was pointed out that in spite of this underlying policy, only 8% of receipts and 9% of the mileage run by the bus fleet represented substituted or direct feeder road/rail services. The balance of 92% of receipts came from independent services, tours, etc. Further analysis broke those figures down into 8% of mileage in respect of substituted rail services; 1% on rail feeder services; 9% on tours and non-scheduled runs; and the remaining 82% on rostered services.

It was noted that the Dublin radial routes were highly profitable, whereas there was a high incidence of uneconomic services in the Monaghan and Donegal areas. Figures included also showed that, between 1944 and 1954, the number of private cars licensed in the counties covered by the GNR road services had trebled.

Several important points were made about how the road service provision was organised within the company. It was stated that road and rail services were bound closely together. This was the result of the basic operational policy decision mentioned above, but was also influenced by how the road services originated within the company.

It was argued that this had several disadvantages for the efficient running of the road operations. It was stated that the effective utilisation of vehicles depended on Station Masters and Agents, but that most of these were wedded to rail and regarded road vehicles with antipathy. However, on the positive side, using existing rail clerical staff on road services allowed these to be administered at a relatively low cost.

With regard to possible future developments, certain recommendations were made. On the policy side, several of these were of a fundamental nature. The following four paragraphs are quoted from the consultants' report:

The GNR(B) should recognise the existence of *greater potentialities for economic and profitable operation in road transport than in rail.* Therefore the existing policy whereby road activities were regarded as secondary to rail should be amended so that road operations would be considered on equal terms to rail.

The Board's policy and intention to provide an efficient transport undertaking, *if necessary predominantly road*, should be made known to all staff, such policy being in the public interest and in the long term in the interest of employees.

The travel requirements of the public should be the criterion for the planning of road services, *rather than the automatic replacement of routes formerly covered by rail.*

Road services on which receipts were in excess of running costs should be encouraged and efforts should be made to retain existing business formerly carried by rail by encouraging its transfer to the Board's road services *prior to the closing of the rail services.* (Author's italics)

On the organisation side, it recommended that a Road Motor Department be established under a General Manager, Road Services, reporting to the Traffic Manager, and that steps be taken to make current railway staff more road conscious, by giving them experience in that function and by the appointment of staff with predominantly road experience to joint posts.

The GNR continued to operate bus services in the Irish Republic up until 1 October 1958, when the Great Northern Railway Board was wound up. The railway network was divided between the UTA and CIÉ. All the remaining GNR buses, 158 in total, were allocated to CIÉ, as the GNR ran no buses north of the border.

Interestingly, to the end, GNR bus drivers were known as 'motormen', in true railway and tramway fashion, rather than as 'drivers'.

The Great Northern continued to operate the buses acquired from the 31 operators it took over, but

GNR Gardner No 324 at Donegal Town. This bus was one of a batch introduced in 1941-42.
Joe Carroll

GNR AEC Regal IV in CIÉ livery. This was one of a batch of 33 buses introduced in 1954-55.

Photobus

also gradually standardised its fleet, mainly on vehicles constructed at its workshops in Dundalk. As noted above, it operated its first bus service, the Drogheda Town Service, in 1929 and by the end of that year the fleet stood at 60 single-deckers.

The first GNR-built bus emerged from the Dundalk works in 1930. This bus was built to the GNR's own design and was fitted with a five-cylinder gas compression engine.

By the end of 1930, a new bus body shop had been erected within the Dundalk works complex. The carriage and wagon building shop was converted into a road motor garage with facilities for carrying out engine repairs, and a paint shop.

The GNR bus fleet initially comprised a mixture of Vulcans, Gilfords, Dennises, Guys, Leylands, Commers, AECs, ADCs and Albions, many of which had been acquired from private operators. A large number of these odds and ends, obtained from operators taken over in Northern Ireland, were later transferred to the NIRTB.

The GNR particularly favoured AECs, both for its bus and later its multi-engined diesel railcar fleets, but it also bought a considerable fleet of Leyland vehicles. The company also built a large number of buses of its own design of chassis and body, but fitted with Gardner engines and proprietary axles and gearboxes. The railway manufactured its own chassis frames, radiators and springs as well as building the bodies.

Ninety-five buses were constructed between 1937 and 1952. The first batch had David Brown gearboxes, but later ones were fitted with Leyland units. This gave strength to the story that what the GNR were building were really Gardner-engined Leyland Lions. The GNR had previously bought petrol-engined Lions.

Building its own buses saved the GNR an import duty of 12½%, imposed on chassis obtained in 'knocked down' form to be assembled locally.

Because the GNR foundry could not initially produce steel castings, an unusual feature of the first batch of buses built were various castings all made of bronze.

Of the various buses acquired second-hand from private operators, some were retained by the GNR for its own use, if they were compatible with standard vehicles in the fleet. These buses included a number of Leyland TS2s bought from HMS Catherwood in Belfast and new to that company between 1928 and 1932.

As mentioned above, at the time of the setting up of the NIRTB, the GNR transferred most of its more modern buses to the south and handed over an assortment of 50 vehicles to the new board. Some of these were buses which the GNR had bought new, namely four Dennis Lancets and 24 Albion PX65s, but 22 were buses it had acquired from private operators when it had taken over their services. These were made up as follows: four Vulcans – two 1927 VWBLs and two Duchesses, dating from 1929 and 1930; six Gilfords – an LL15, an OT (both 1928), a GL, a 15SD and two 1660Ts (all 1929); eight Dennises – three Gs (1928) and five Es (1927-29); two Guy ONDs of 1929-30; and two 1931 Commers – an NF6 and a 6TK.

During 1941-42 the GNR hired 20 vehicles to the NIRTB, which was left short of buses due to having had vehicles requisitioned by the military. Eighteen of these hired buses were Leylands and the other two were AEC Reliances.

All 50 ex-GNR buses were withdrawn by the NIRTB between 1935 and 1944 and the last GNR-built bus in passenger service was withdrawn by CIÉ in 1973.

The GNR bus livery was Oxford blue and cream, with black mudguards and cream wheels. The words 'Great Northern' were carried on the waistband. The fleet number was carried on the front and back and on the sides low down, near the front, on some vehicles. On earlier liveries the 'GNR' monogram was carried centrally on the side. In later years, this was dispensed with and the sides remained unadorned, but in the final livery style the full coat of arms was carried on the sides.

The SLNCR

As with the Donegal operators, the poor state of the roads in the SLNCR's area protected it from the beginnings of the road competition being experienced by railways in more prosperous parts of the country, after World War I. However, by the late 1920s, the company was starting to look at the potential threat that road vehicles posed to its position and how it might deal with this problem. As with the GNR, CDR and LLSR, the SLNCR had to cope with the additional inconveniences of having to operate over the international border, created when Ireland was

divided politically in 1921.

In 1928, the Sligo and Leitrim considered becoming a road operator in a small way and looked into the possibility of buying a bus to provide a service to Kilmakerrill, from the company's headquarters at Manorhamilton. However, it did not follow through with this idea.

In 1933 the Central Omnibus Company, set up and run by a Mr Appleby, began operating between Sligo and Enniskillen. These were the terminal points of the SLNCR system and so the Central Omnibus Company entered into direct competition with the railway for its passenger traffic. The railway company made an approach to Appleby with a view to taking over his company.

However, Sir Felix Pole's inquiry into the whole relationship between road and rail travel in Northern Ireland was in progress at the time. This created a degree of uncertainty and, because the SLNCR was about to introduce its own railbuses, it decided to hold off on the Appleby take-over.

As it turned out, this was a good move since, as has been seen, Pole's inquiry led to the establishment of the NIRTB. The board absorbed the 11 mile stretch of Appleby's operation between Enniskillen and the

Sligo Leitrim and Northern Counties Railway Commer Commando at Manorhamilton.   JC Gillham

border, leaving him with the 30 mile section between the border village of Blacklion and Sligo.

Nothing more happened with regard to establishing a bus service until 1944, when Appleby was approached again with a view to selling his operation. He agreed to do so and in early 1945 the SLNCR took over his two buses and the Blacklion to Sligo service. Between 1945 and 1949, the SLNCR developed its bus services by operating additional routes to that which they inherited from Appleby and at its height required the use of four vehicles to provide this service.

With regard to the buses themselves, the two taken over from Appleby were Bedford WLB single-decker petrol-engined vehicles – a 20-seater built in 1932 and a 26-seater, built in 1938. To this were added, in 1946, two ex-GNR petrol-engined single-deckers. These were ex-HMS Catherwood/ex-GNR Leyland TS2 31-seater No 296, built in 1928 and re-bodied and renumbered No 129 in 1937; and ex-GNR Leyland TS2 No 96, built in 1931 and re-bodied in 1938.

In 1948, the Appleby buses were replaced with two 32-seater diesel-engined Commer Commandos. The ex-GNR buses were re-engined as diesels in the same year. In 1952, the two original GNR Leylands were replaced by two further second-hand 35-seater GNR-Gardner diesel-engined vehicles, GNR fleet numbers 214 and 218, both built in 1937. GNR No 214 lasted with the Sligo and Leitrim until 1956, when it was replaced by a similar ex-GNR bus, No 208, also built in 1937.

Between 1945 and 1957, the SLNCR used nine buses though, as noted above, only four were in service at any one time. When the SLNCR closed down in 1957, both its road and rail services were discontinued, the road passenger and freight services being replaced by CIÉ using its own vehicles. None of the SLNCR buses were taken over by CIÉ.

The bus livery was similar to that applied to the railbuses and railcar B – regent green below the waist, with olive green above, separated by one or two narrow black lines. The company's initials were applied along the narrow waist panel below the windows.

Under state ownership

CIÉ buses

Because of the complexity of the development of CIÉ's bus operations since its foundation, only a general overview of the main types of buses introduced can be given here. For more detail, the reader should consult Michael Corcoran and Gary Manahan's excellent book *Winged Wheel, CIÉ Buses 1947-1987* and also the PSV Circle/Omnibus Society/TMSI's *CIÉ fleet list PI 2*.

This survey will cover the period from 1945, when the integrated road-rail CIÉ organisation was established, until 1987, when CIÉ was split into three separate operating companies – Iarnród Éireann, Bus Éireann and Bus Átha Cliath.

The government believed that the pre-war position of transport in the country had been unsatisfactory and that, outside Dublin, public transport facilities were poor. It had also been disappointed with the results of the 1932 and 1933 Transport Acts. The solution was to merge the GSR and the DUTC into a single integrated transport company – CIÉ.

CIÉ was set up on 1 January 1945. It was not to be a nationalised concern, because the Irish government believed that nationalised transport almost inevitably had to be subsidised and that subsidies were not necessary. It was thought that public transport could be financially viable.

The Dublin United Tramways Company had been incorporated in 1896 and had eventually developed a route network of about 60 miles, over which it operated 330 trams. In 1925 it diversified into operating buses and in 1934 acquired the General Omnibus Company, which brought to the DUTC a flavour of London Transport. The GOC had modelled its operation on London Transport and this was reflected in the subsequent design of the DUTC badge, introduced in 1941.

This badge, which was later adopted in a modified form by CIÉ as the 'Winged Wheel' (but more commonly, if irreverently, known as the 'Flying Snail') was clearly based on the London Transport 'Bar and Circle' logo. The DUTC version had the words 'Iomchar Átha Cliath' (Dublin Transport) on

Ex-DUTC four-cyl AEC Regal II in CIÉ livery on Dublin Quays. This was one of a batch introduced in 1933-36.

Photobus

the centre bar.

The DUTC began tramway abandonment in 1938. By 1941 about 220 trams and some single-decker buses had been replaced by 242 Leyland Titan double-decker buses. Further deliveries were stopped by wartime conditions, so the trams survived into CIÉ days, final closure not taking place until 1949. However, the company changed its name to 'Dublin United Transport Co' in 1941.

In the early 1930s the DUTC had bought a mixture of Leylands, AECs and Dennises. When the General Omnibus Company was taken over in 1934, it acquired a fair number of Albions but, by 1944, it had standardised on Leylands and AECs.

The livery was originally ultramarine and ivory. This was changed to grey and white in 1929 and in 1935 was changed again to Audley green and cream, with olive green roofs. In 1941 it adopted a livery of olive green and eau-de-nil.

At its foundation, CIÉ owned 618 serviceable buses and 113 trams. All the trams and 364 of the buses came from the DUTC and the other 254 buses from the GSR. CIÉ would provide about 80% of the state's public transport services and of the remaining 900 independent companies, licensed to carry 'for hire or reward', most were one-vehicle concerns and

only 28 were bus companies.

In January 1945, CIÉ took over two quite different bus operators. Although, on the face of it, CIÉ ran a unified road transport service, using the same numbering series and livery, it actually ran two bus operations, one corresponding roughly to those of the DUTC and the other to the GSR. Dublin city and provincial services were always clearly distinguished.

This was clearly underscored in 1987 when the two new operating companies were set up. Bus Éireann, the provincial bus service provider and Bus Átha Cliath, the Dublin city service, seemed to be the IOC and DUTC reborn, even down to the liveries chosen initially by the two companies.

When CIÉ was set up, a Rolling Stock Engineer (Road) was placed in charge of all the company's road vehicles. All major mechanical work was carried out at the GSR's former workshops at Broadstone. Body-building was carried out at the DUTC's Inchicore works, called Spa Road to distinguish it from the railway works.

The fleets taken over from the previous two companies comprised 254 serviceable buses from the GSR and 364 from the DUTC. The GSR fleet consisted of 220 single-deckers and 34 double-deckers. The single-deckers were mainly Leylands –

Ex-GSR Leyland TD4 fitted with later CIÉ body. The original body was a low bridge type.   Photobus

Cubs, Lions and Tigers. There were also three AEC Regal single-deckers and 34 Leyland Titan double-deckers. The bulk of this fleet, 229 vehicles in all, were petrol engined.

The 364-strong DUTC fleet consisted mainly of Leylands – 84 Lion single-deckers and 242 Titan and four Tiger double-deckers. The fleet also included 34 AEC Regal single-deckers. All the DUTC fleet was diesel-engined.

Withdrawal of ex-DUTC buses began in 1950. The bulk of the single-deckers were gone by 1954, but some of the double-deckers survived until 1960. Most of the ex-GSR fleet, with its uneconomical petrol engines, were gone by 1954. However, the last of its 14 diesel-engined Titans was withdrawn as late as 1960.

A large scale bus building programme was begun in 1946. Twenty-seven double-decker chassis were purchased. These were made up of one AEC Regent, six Daimler CWD6s and 20 Leyland OPD1s. These chassis were imported in knocked down form, assembled at Broadstone and then bodied at Spa Road. Although, these bodies are described in the fleet listings as CIÉ, they were in fact Leyland or Alexander designs.

The first new single-deckers were on 15 GSR Leyland TS11 Tiger chassis, which had been in store throughout the war and were bodied in 1947. In the same year, ten more AEC Regent double-deckers were built. These were to be the last AECs to be bought by CIÉ for some years.

One of the first acts of the new CIÉ board was to sign an agreement with Leyland in 1947, under which CIÉ agreed to buy only Leyland premium bus chassis for 25 years and, in return, Leyland would assist CIÉ to set up and operate a chassis building plant. In the event this plant was never built.

In 1948 the very successful Leyland OPD2 66-seater double-deckers fitted with the 9.8 litre Leyland 0.600 engine were introduced. This remarkably reliable design of engine remained in production for nearly 25 years. It was the standard engine fitted to the CIÉ bus fleet for the next 15 years. In addition to these 'Spa Road standards', CIÉ also imported 43 Leyland PD2/3s with Leyland bodies, the 'Bolton' class and, a year later, 50 more Leyland bodied 'Capetown' class Leyland PD2/3 double-deckers. These 58-seaters were used to replace the last of Dublin's tram services.

Between 1949 and 1951, 40 P-class single-decker OPS3 Tiger chassis were acquired and bodied by CIÉ for tours work. Thirty were fitted with full-fronted coach bodies. Twenty of these 30-seater coaches were given the names of Irish rivers. This batch of coaches formed the basis of what would become the very successful tours business. In 1949 CIÉ also bought six Bedford BP class petrol-engined coaches for airport work. These remained non-standard vehicles for the company.

Deliveries in 1952 and 1953 were dominated by half-cab rear-entrance Leyland Tigers, even though CIÉ had experimented with underfloor-engined, front door entrance Leyland Royal Tiger single-deckers for tours work. In 1954, 38 underfloor-engined 45-seater U-class Royal Tigers were introduced for stage carriage work but, as they were also intended for crew operation, they were fitted with rear entrances.

By the mid 1950s, demand for bus services was expanding faster than CIÉ could handle it and the bus fleet had increased to over 1,000 buses and coaches. Dublin stage carriage work accounted for about 600 vehicles, 350 on other stage carriage work and the rest on tours. Double-deckers had been introduced

CIÉ standard Leyland OPD 2/10. This bus was one of a batch built in 1957-58.   Photobus

into Cork, Galway, Waterford and Limerick and even some heavily trafficked country routes. This situation did not change until long wheelbase single-deckers became available.

In 1958 a big change hit CIÉ with the absorption of the GNRB road fleet. As already noted, the GNR operated services north of a line between Dublin and Sligo, including north-east Dublin suburban services.

By the time the GNRB was wound up, it had a fleet of 158 buses. The single-deckers comprised 55 five-cylinder GNR Gardners, four Leyland Royal Tigers, 30 AEC Regal IIIs and 33 AEC Regal IVs. In addition there were 31 AEC Regent and five ex-DUTC Leyland TD5 double-deckers.

The GNR used its double-deckers on the Sligo-Bundoran route, the Newry-Dundalk cross-border service and on the Dundalk and Drogheda town services. Under CIÉ most of the ex-GNR fleet remained in its original operating area.

This period saw the closure of many railway lines – the GNR and SLNCR standard gauge lines and the County Donegal and Cavan and Leitrim narrow gauge systems. Many of the replacement cross-border

bus services were operated jointly by CIÉ and the UTA. As a result of the need to provide services to replace those once provided by the closed railways, CIÉ's fleet was expanded to over 1,300 buses.

Between 1959 and 1961, CIÉ introduced a fleet of 30'0" double-deck buses, the RA class. These 74-seater vehicles were based on the Leyland PD3/2 chassis and were fitted with semi-automatic transmission and air brakes. They were used to replace the last of the pre-war crash gearbox double-deckers. In 1961 CIÉ ordered three AEC Regent double-deckers, which were used to replace the Waterford to Tramore train service.

Another famous Leyland model, the Leopard, made its appearance in 1960. The version introduced by CIÉ was the L2 and was fitted with the legendary 0.600 engine. CIÉ took delivery of 170 of these buses between 1961 and 1964. The Leopard proved to be an extremely economical and reliable vehicle and in Ireland the model gave many years excellent service with CIÉ, the UTA and Ulsterbus.

On the tours front, in 1960, CIÉ introduced a small number of new coaches based on the Royal

*CIÉ Worldmaster coach WTS
'Briomhar'.*   Photobus

Tiger Worldmaster chassis. Twenty-three of these coaches were bought, plus 13 PSU series Leopards which complied with Northern Ireland PSV regulations, unlike the Worldmaster. These 40-seater buses were 36'0" long and were fitted with Ogle-designed bodies. In 1962 they were painted in a cream and brown livery. This was later applied to the new express coaches introduced on new services from 1965.

By the early 1960s CIÉ was experiencing a shortage of double-deckers, but the model now being offered by Leyland was the Atlantean. A demonstrator was tried in Dublin in 1960, but CIÉ was not impressed and no orders followed at that stage. This demonstrator had previously been on trial with the UTA and while in Dublin retained its UTA livery. The UTA did not order any Atlanteans either. The demonstrator was eventually sold to the Lough Swilly, becoming their fleet number 87.

CIÉ briefly toyed with introducing the Daimler Fleetline which had become the very successful standard double-decker with Belfast Corporation, but eventually chose the Atlantean, their first order comprising 341 vehicles. Between 1966 and 1974 CIÉ bought 602 of these buses.

The Atlantean design turned out to be an unwise investment. The design was complicated, requiring extra maintenance, and the initial vehicles, fitted with the 0600 engine, were underpowered. Later buses were fitted with the more powerful 0680 engine, but the Atlantean was never as reliable as previous Leyland designs. Ulsterbus, which introduced the Atlantean in 1971, also found it unreliable. After many years of excellent PSV designs from Leyland, the Atlantean turned out to be a dud and its unpopularity must have been a contributory factor in the eventual demise of the company.

In 1971 the next problem product from Leyland made its appearance – the M class Leyland Leopard. CIÉ had begun to introduce one-man-operation at this time and 213 of these 40'0" 55-seaters were built for express and long distance work. However, the buses displayed many of the problems being experienced with the Atlanteans. As a result, CIÉ began to replace Leyland engines with power units manufactured by Cummins, General Motors and DAF.

However, CIÉ did manage to avoid buying another poor Leyland design – the Leyland National integral-bodied single-decker. Like Ulsterbus in the north, CIÉ only acquired one of these buses, a demonstrator which it tried out during 1975-76. It was not a success. CIÉ decided on a different strategy for future bus supply.

In 1973 Van Hool-McArdle took over the Spa Road works from CIÉ and began to bring out new designs. In the event, the Van-Hool project failed and CIÉ had to start all over again to plan for future bus provision.

Their next attempt, in 1980, involved a new factory in Shannon operated by the Canadian company Bombardier. During that period it built 365 double-deckers, 201 city service single-deckers, 51 provincial service buses and coaches and 225 rural service single-deckers for CIÉ. There were problems with this plant too and it only lasted six years.

After the failure of the Bombardier project, CIÉ decided in future to buy proprietary designs and to give up attempting to have buses built in an Irish factory dedicated to producing their vehicles. Recent double and single-deckers, based on Leyland and Volvo chassis, have been built for Dublin Bus in Northern Ireland by the Scottish bus builder Alexander, in their factory on the outskirts of Belfast, and by Wright of Ballymena.

CIÉ attempted to achieve maximum integration between its road and rail provisions in both urban and long distance services. On long haul routes, even where the Expressway bus services parallel those of the railway, the appearance of direct competition is misleading. Mainline trains now operate with fewer intermediate stops and thus some places find that the fall in their rail services is partly compensated for by the provision of the express bus service. Thus, CIÉ's long distance bus services are designed to complement its railway provision.

Where competition did exist between bus and rail, it was between rail and private bus operators, who also competed with CIÉ's express bus services. On occasions when CIÉ tried to provide feeder bus services to the railways, private operators often responded by providing competing point-to-point direct services.

The best example of urban bus-rail integration is to be found in Dublin. The Dublin Area Rapid Transit system, or DART, began operating in July 1984 and from the outset it was popular with the travelling public. Because of industrial problems, it was another 18 months before the feeder buses, in their DART livery, commenced services. However, when the service finally began, two million additional passengers were carried by the DART in its first year.

CIÉ also provides a Dublin station-to-station bus link. A regular service of single-decker buses links Connolly and Heuston stations, directly from each terminus.

As noted above, in 1987 CIÉ was re-organised as a holding company with three operating subsidiaries – Iarnród Éireann, Bus Éireann and Bus Átha Cliath. From that point onwards the operating subsidiaries had a greater degree of independence, but the integrated road-rail nature of CIÉ will continue to be fostered.

UTA buses

Unlike other companies surveyed so far, the UTA was not, strictly speaking, a railway company which introduced bus services. The Authority was the result of a marriage forced on the Northern Ireland Road Transport Board and the ailing railway companies by a Northern Ireland government desperate to find some solution to the transport problems which had plagued it almost since the state was set up, in 1921. In the final analysis, the UTA turned out to be essentially a road operator, saddled with railways it didn't really want. Where possible, it closed and

replaced the railways with road services, to such an extent that cynics said that the initials 'UTA' stood for 'Ulster Track Abandonment'!

It is proposed here to deal with the UTA's bus operations in general terms, particularly where they impinged on those of the railways that the UTA absorbed. In broad terms, it is intended to look at the types of buses that the Authority used often in competition with, or even replacement of, its own rail services. For more detail on the UTA bus fleet the reader is referred to the PSV Circle's *Fleet List P15*, covering the NIRTB, UTA and the early years of Ulsterbus, or Brian Boyle's *Buses in Ulster Vol 2, The Ulster Transport Authority 1948-67* (Colourpoint, 2000).

Turning now to the bus fleet situation, when the NIRTB began operations, in October 1935, it took over the vehicles owned by the various bus operators in the province. These comprised vehicles from the 'big five' bus operators – namely, 169 buses from the Belfast Omnibus Company, 70 vehicles from HMS Catherwood Ltd, 14 from the BCDR, 50 from the northern section of the GNR and 131 from the LMS (NCC). Later in the year it absorbed 182 further buses from 40 smaller operators and, the following year, 71 buses from 17 private hire and tour operators. In total, this meant that the board started life with 687 buses, of 27 makes and 97 types.

The NIRTB started to rationalise its fleet as a matter of priority and immediately ordered 121 new buses. This order consisted of 15 diesel-engined Leyland Titan TD4 double-deckers, 20 Leyland LT7s, 40 diesel AEC Regals, 42 Dennis Lancet I and four Dennis Lancet II single-deckers. By the outbreak of war, in 1939, the fleet had been reduced to 478 buses and of the 434 buses taken over by the board in 1935 from the 'big five', 218 were already gone.

The only new buses obtained during the war were 175 utility Bedford OWBs, one Leyland TS11, ten Leyland TD7s, three Dennis Lancet II single-deckers and seven Guy Arab and two Bristol K5G double-deckers. By the end of the war, the fleet was badly run down.

In 1946 the first post-war buses began to arrive. These were 20 AEC Regals and seven Leyland Tiger PS1 single-deckers with Duple bodywork, to the board's specification and design.

The NIRTB intended to become self-sufficient in

NIRTB Leyland PD1 Z907 in UTA livery, with final UTA coat of arms.   Photobus

bus body building, so a former aircraft factory at Dunmurry outside Belfast was taken over and the board began to build its own buses, to a standard design, on Leyland Tiger PS1 and PS2 single-decker chassis. A limited number of double-decker buses was also built.

As a result, when the UTA was established in 1948, it was able to continue a fleet standardisation programme that was already in place. Only the bus livery changed. The two shades of green used by the NIRTB on waist and lower panels, which had been adopted from the BOC, were reversed in the UTA livery.

The standard UTA vehicles were the Leyland PS1 and PS2 single-decker fitted with the NIRTB designed 34-seater body, and the Leyland Titan PD1 and PD2 fitted with the board's low bridge 53-seater double-deck body.

In 1949 the Authority opened a new bus-building facility in Duncrue Street, Belfast, in part of the ex-NCC railway workshop premises there. However, the Dunmurry facility was not finally closed until 1951. The output of the UTA's body-building plants was about 150 buses per year. This allowed the

withdrawal and replacement of many of the remaining pre-war and all the war-time utility buses.

In 1951 the UTA switched to using higher capacity underfloor-engined vehicles, possibly with an eye to future one-man-operation, and 61 Leyland Royal Tiger 42-seater PSU1/11 single-deckers were fitted with front entrance bodies of UTA design. Between 1956 and 1964, 306 further single-decker front entrance vehicles were constructed on Leyland Tiger Cub PSUC1/5, AEC Reliance and Albion Aberdonian MR11L chassis. The final single-decker buses bodied by the UTA were 50 school Bedford SB5s, with 51-seats, in 1964.

In 1948 the UTA had inherited 74 low bridge double-deckers from the NIRTB. These were a mixture of Leyland TD 1 – 7s, Leyland PD1A and PD2s, Guy Arabs and AEC Regents. Some of them dated back to BOC and Catherwood ownership.

The bodies on the Leyland 'low bridge' buses had the interesting feature that the seating on the upper-deck consisted of four-seater benches, with a sunken access aisle along the right hand side. It was this feature which allowed the buses to be built low enough to negotiate the many low railway bridges in

Right: UTA Leyland PS2/1 C8817.
WH Montgomery

Below: UTA Leyland PSUC1/S Tiger Cub in UTA two-tone green livery.
Photobus

Bottom: UTA AEC Reliance in UTA two-tone blue livery.
Photobus

the province. However, it meant that the passengers sitting in the right-hand window seats downstairs had to be careful of their heads when getting up!

The UTA continued to build this type of body, mounted on the PD1A and PD2 chassis. Originally a 53-seater, a longer 59-seat version was introduced when the legal limits on bus sizes were changed in 1951. Fifty-one of these 'low bridge' double-deckers went into service between 1950 and 1955.

In 1948 a 27-seater prototype vehicle was built at the NIRTB's Dunmurry works, as a semi-decker or half-decker with a large luggage compartment. This vehicle was finished in a variation of the Authority's livery which was close to the second NIRTB colour scheme.

This experiment in vehicle design is in itself interesting. During World War II and the post-war period, parcel traffic carried by buses increased greatly and it was decided to introduce vehicles capable of dealing with it on trunk routes, particularly to Enniskillen and Omagh. Six PS2/10 chassis were initially ordered from Leyland. This was the PS2 chassis with a 'dummy' third axle inserted immediately behind the front axle, to make the chassis a six-wheeler. It allowed a 30'0" length body to be fitted, instead of the normal 27'6" one.

However, the authorities would not pass either the Enniskillen or Omagh

UTA Leyland PD2/1, with 'low bridge' body, at Smithfield depot, Belfast.
WH Montgomery

routes for its use, so the semi-decker idea had to be abandoned. One other chassis had been delivered to the UTA, but the other four were quickly cancelled. The first semi-decker was followed, in 1951, by a second similar vehicle utilising the second chassis. The prototype vehicle spent some years in Ballynahinch, but generally the two vehicles were used on the Belfast airport services because of their extensive luggage space and their full potential was never fully exploited.

In 1958 these semi-deckers were stripped of their bodies and their chassis converted to PD2/10cs. They were fitted with the standard UTA/MCW 60-seat 'high bridge' double-deck bodies. In this form one of them is now preserved.

During the 1950s there was a great increase in the use of double-deckers to provide economies because of falling traffic. The Authority had 156 Leyland Tiger PS2s stripped of their single-deck bodies, their chassis strengthened and then fitted with high bridge 60-seat double-deck bodies utilising Metro-Cammell-Weymann (MCW) frames.

These buses entered service between 1956 and 1958. Because they were built on what were reconditioned, front-engined, single-decker chassis, they were light for double-decked vehicles. When lightly loaded they tended to bounce, even on good roads, and so were very uncomfortable for passengers, particularly those seated upstairs.

In 1957 the operation of the Erne Bus Service of Enniskillen was acquired. This operator had not been compulsorily taken over in 1935, because it was a cross-border operator. To avoid any legal or political problems, the UTA set up a subsidiary company called the Devenish Carriage and Wagon Co Ltd to operate these services.

With this acquisition, came the eight buses of the Erne fleet. These consisted mainly of Leylands, namely one LT4, one LZ1, two CP01s, two PSU1/9s, and one PSU1/12. In addition, there was one Magee-bodied Dodge SBF. On the Leylands there was a variety of bodywork – Magee, Murphy, Saunders-Roe and Burlingham. Five of the eight were withdrawn by 1960, but the three PSU1s lasted until the end of the UTA.

One of the Erne buses, a Leyland LT4, had an interesting pedigree. This bus had been built in 1932 for the Irish Omnibus Co, as their fleet number 730. It had then become part of the GSR fleet, before being sold by them to the Erne Bus Co. It was withdrawn

UTA PD2/10C No L725 with MCW 'high bridge' body at Ormeau Avenue, Belfast.
WH Montgomery

by the UTA in 1960.

In the period between 1959 and 1963, the UTA built 142 Leyland Titan PD3s. Like the PD2s, these were a high bridge design on MCW frames, but with a full width cab enclosing the engine. They were fitted with a forward staircase and an air-operated sliding door sited just behind the cab. Seating was 67 on earlier models and 69 on later, though a few were converted to 64-seat versions with extra large boots for airport work. Most of these buses were PD3/4s, but five of the 67-seaters, fleet numbers N986 to N990, were PD3/5s. These latter vehicles were fitted with gearboxes which could be operated in either semi-automatic or fully automatic mode.

By the end of the UTA period, there were 356 double-deckers in the fleet.

The NIRTB had no coaches and used buses for private hire and tour work. In 1951 the UTA built two 36-seat coaches, Nos E8930 and E8931 on the Leyland PSU1/11 chassis. They also acquired coaches from Southdown, Ribble, Barton and Wallace-Arnold to operate those companies' tours, as British operators were not allowed to operate in Northern Ireland in their own right. Coach-type, dual

purpose buses were also built, such as the 1960-built Albion Aberdonian MR11Ls, the 1962-build Leyland Tiger Cub PSUC1/12s and the 1963-build AEC Reliances. These were all fitted with coach seats in what the UTA called a 'semi-luxury' type body.

A fleet of 35 Bedford SB5s, with 41 seats and Duple bodywork, delivered in 'knocked down' condition, was built by the UTA in 1964 for 'Hotel Tours' contracts. They were painted in the liveries of the various hotels for which they worked on contract.

In 1965 the UTA introduced four 38-seater Leyland Leopard PSU3/3RTs, with UTA bodywork, for use on Express Services using the new M1 motorway which was being built from Belfast towards Dungannon. In 1966 two more were added, but with 37 seats. They were the only coaches in the fleet to be fitted with toilets, apart from the two 1951-built coaches E8930 and E8931. These new coaches were marketed under the brand name 'Wolfhound' and were the UTA's flagship vehicles. They were supposedly inspired by the Greyhound buses seen by a top UTA official when on a visit to the USA.

By 1966 the UTA's days were numbered and the last major batchs of vehicles to enter the fleet were 45

UTA Leyland PD3/4 Nos Q826 and P888 at Oxford Street depot in Belfast. This bus station was built on the site of the original Belfast Central Railway's Central station.   Photobus second-hand ex-Ribble Leyland Royal Tiger PSU1/13s, with Leyland bodywork; 48 ex-Edinburgh Corporation Leyland Tiger Cub PSUC1/3, with MCW bodies; and 15 ex-Western SMT Leyland Leopard L1 coaches, with Alexander bodywork.

On the formation of Ulsterbus the following year, these were earmarked to replace the ageing Leyland PS1 34-seater half-cab single-deckers, which were unsuitable for conversion to one-man operation.

The railway services, closed by the UTA in the 1950s, were replaced in the main with the standard NIRTB/UTA 34-seater single-decker and 53-seater double-decker buses. Initially, these operated to the railway timetable, sometimes starting from the Belfast main line stations from which the train services had been withdrawn, and often terminating at the railway stations at the other end of the closed line.

Typical of the scenes which must have occurred all over Northern Ireland during the period of rail closures, is the activity recorded at Donaghadee on the morning of 24 April 1950, the first day of working without train services on the BCDR main line. That day an ex-BOC/NIRTB AEC Regent of the UTA, two standard ex-NIRTB 34-seater single-deckers and another unidentified, but more modern, low bridge double-decker were all lined up at the railway station to operate the train replacement service. It is worth noting that on that date the AEC Regent was 18 years old, so it is probable that everything that could move was out that day to make sure the transition from rail to road went smoothly!

Many of the old railway stations were later converted to bus stations, but retained their obvious railway characteristics for many years after the railway was closed, eg Newcastle, Downpatrick and Ballynahinch on the BCDR, Armagh on the Great Northern and Limavady and Cookstown on the NCC system.

Moreover, the UTA did institute through-ticketing facilities and published a common road-rail timetable, showing all the connections between buses and trains. Joint road-rail termini also provided for convenient road-rail interchanges.

The 'Wolfhound' motorway express coaches introduced in 1965 and 1966 were designed to replace, in part, the train services withdrawn in 1965. In the 1966 timetable these buses were booked one hour, 17 minutes from Belfast, Great Victoria Street

(from where the trains on the closed service used to leave) to Dungannon, non-stop, and a further 48 minutes to Omagh, with only one intermediate stop at Ballygawley. It must be mentioned though, that at this time the motorway only extended as far as Lurgan. Dungannon was not reached until the mid 1970s.

This was comparable to the timing of a *stopping* train on the GNR 1928 timetable, leaving Great Victoria Street. This was booked for one hour, 18 minutes to Dungannon, making six stops, and a further 46 minutes to Omagh, making five intermediate stops. However, the GNR *express* train, complete with buffet car, was booked one hour, six minutes to Dungannon, with three stops, and a further 40 minutes to Omagh, non-stop.

Buses replacing rail services did not fare so well on those parts of the UTA not favoured by the use of motorway express coaches. If, for example, the Belfast-Downpatrick BCDR train service of 1938 is compared with the bus service in the 1966 UTA timetable, it can be seen that, whereas the stopping train took 55 minutes for the trip in 1938, the ordinary service bus in 1966 took 70 minutes, following a shorter route. On the plus side there was, in 1966, a much more frequent service of buses to Downpatrick. There were 19 bus departures on the Monday to Friday schedule, as against eight trains in 1938.

The BCDR Belfast to Newcastle stopping train took one hour, 12 minutes for the trip in 1938, whereas the bus took one hour, 33 minutes in 1966, again following a shorter route. However, once again, there was an increased frequency of buses – 14 Monday to Friday in 1966, as against six trains in 1938.

The new fleet of UTA semi-luxury coaches, which started to appear from 1960, were used to improve the quality of travel on Limited Stop and Express services, often on routes from which rail services had been withdrawn, as well as on tour and private hire work.

The livery of the NIRTB was originally overall dark green, with light green narrow bands at waist level. This was later changed to a light green livery, with a dark green waistband and medium green mudguards. Window surrounds and roof were green-cream. Wheels were painted in matt khaki. The words 'Northern Ireland Transport' (in capitals) appeared in gold along the waistband. For a short time a London

UTA Leyland PSU3/3RT Leopard for use on the 'Wolfhound' motorway express service.   WH Montgomery

Transport-style wheel and bar symbol was carried on the side. This was removed when London Transport objected. The fleet number and year identification letter appeared inside a circle about 9" in diameter, on the front offside and rear nearside of the bus. Circle, number and letter were yellow.

As noted elsewhere, the original UTA livery style was identical to that of the NIRTB, except that the dark and light green colours were reversed. A new logo was also introduced, which consisted of the red hand of Ulster on a white shield, placed inside double gold coloured circles. The words 'Ulster Transport' were inscribed in gold lettering between the two circles. The fleet numbering system, method of display and positioning remained unchanged from that of the NIRTB.

In 1951 a light blue-green, with dark green relief was used on two new luxury coaches, and variations on that theme were used afterwards on all coaches and dual-purpose buses. In 1959 an insipid colour scheme was introduced. This consisted of a light blue-green without relief, except for the greenish-cream window surrounds and roof. No lining was applied. This was used on the new batch of PD3 double-deckers and on all subsequent new stock, although most of the older vehicles remained in their original colours until the end of the UTA.

In 1960 a new heraldic coat of arms, complete with Latin motto, 'Transportatio Cultum Significat',

was introduced to replace the 'red hand' symbol. It appeared on all new vehicles and was applied to most of the older road and rail stock, before the Authority was dissolved.

In 1961 a livery of Elizabethan blue and hyacinth blue was introduced for coaches and, in 1965-66, a livery of grey and white, with a red flash, was adopted for the 'Wolfhound' express coaches. These carried the word 'Wolfhound', in capitals, on the front and a picture of a leaping wolfhound on the sides. The new Ulster Transport coat of arms appeared on the centre of the front of the vehicle, below the windscreen.

In the early years of the UTA three Leyland PS1s were converted to 29-seaters and painted in the British European Airways livery. They operated the original airport bus service between Great Victoria Street rail/bus station and Belfast's airport at Nutts Corner.

As regards the final batches of second-hand buses bought by the Authority in 1966, only a few were repainted into UTA colours. Most ran in their original liveries, until painted into Ulsterbus blue and white after 1967.

Northern Ireland Railways

Northern Ireland Railways took over the railway operations of the Ulster Transport Authority in April 1968, under the provisions of the Transport Act (NI)

1967. The railway division of the UTA had traded between July 1966 and April 1968 as 'Ulster Transport Railways', prior to its establishment as a private limited company, although 'NIR' began to appear on repainted rolling stock from mid-1967. NIR inherited a network which was not in the best of shape, having suffered, during the UTA's final years, from under-investment and minimal levels of maintenance.

In Belfast it inherited three termini, all of which had been built in more prosperous times by the independent railway companies, and on the grand scale. For the greatly reduced number of destinations now available to rail travellers, these stations were all too big and badly run down. All were to suffer, in the late 1960s and early 1970s, from the effects of terrorist attacks.

Another problem was that the three termini were quite far apart and such direct rail links as had existed between them in the past, had been severed by the UTA. Previously, the Belfast Central Railway had connected the GNR with the BCDR terminus and the harbour railway network had connected the NCC with the Belfast Central Railway. The severing of these links had the effect of requiring the railway to operate three sets of engineering facilities – one at Queen's Quay to service the rolling stock on the isolated Bangor line, the others at York Road and Great Victoria Street to serve the rest of the system.

Another factor was that only one station, the ex-GNR terminus in Great Victoria Street, was anywhere near Belfast's central business district. Of the other two, the ex-BCDR terminus at Queen's Quay was on the 'wrong' side of the River Lagan and, like the ex-NCC terminus at York Road, was a good 20 minutes walk from the city centre.

In 1972, the Northern Ireland government decided that a major investment programme in the rationalisation and integration of the Belfast rail network would take place. Under this plan, Queen's Quay and Great Victoria Street stations would be closed and replaced with a new 'Central' station, at East Bridge Street. The new station would be built on the site of the old railway cattle handling area at Maysfield, on the eastern edge of the city centre.

As part of this plan the original Belfast Central Railway link would be re-opened, with a new halt built at Botanic Avenue, close to Queen's University and the southern edge of the central business district. York Road station would remain to serve the Larne line. All Derry-Belfast services would leave the ex-NCC main line at Antrim and make a detour to Belfast, over the Antrim to Knockmore Junction branch, and thence to Central station along the ex-GNR main line and former Belfast Central Railway. The Antrim-Knockmore goods-only line was to be completely re-laid and re-signalled, and most of the long-closed stations along it re-opened.

Although this scheme was a great boost to NIR, and expressed a level of official confidence in the future of the railways, it resulted in two problems in Belfast. Firstly, the closure of Great Victoria Street station deprived NIR of its only real city centre station, as the new 'Central' station was anything but central in its location. The Great Victoria Street station site, and its approaches, were required by the Department of the Environment for a link road between Great Victoria Street and the proposed elevated motorway to link the M1 and M2. A scaled down version of this motorway would eventually become the 'Westlink'. The direct road link with Great Victoria Street was never built.

To be fair, Central station was to have been one part of a major

Ex-Edinburgh Corporation Ulsterbus No 1011, a Leyland PSUC1/3 with Weymann body, approaching Central station. It is wearing Citybus and Citylink stickers.

Author's collection

Right: *Citybus Bristol RELL in NIR Rail-Link livery at York Road station.* RC Ludgate, Author's collection

Below: *NIR Suburban sector liveried Citybus Leyland Lynx at Short Strand depot.* Author

Right: *Citybus Bristol RELL in NIR Rail-Link livery at York Road station.*
RC Ludgate, Author's collection

Below: *NIR Suburban sector liveried Citybus Leyland Lynx at Short Strand depot.* Author

commercial redevelopment scheme in that part of the city which, in the event, has only begun to happen in the late 1990s. Failure to develop the city centre as planned in the 1970s, left Central station out on a limb, with the most convenient station to the city centre being the new halt at Botanic Avenue. This halt had to be rapidly expanded in size almost as soon as it opened.

The other problem was that York Road station was not part of this railway link up and, because of urban redevelopment and motorway building plans, it was left stranded in splendid isolation on the extreme edge of the city centre. For passengers arriving at either York Road or Central, scheduled bus services would have to be relied upon and, particularly in the evening, these bus services were unsatisfactory.

Because of this potential problem, for the opening of the new Central station, in April 1976, NIR and Citybus Ltd agreed to set up the 'Citylink' service.

As originally constituted, the 'Citylink' service ran a circular route and only linked Central station with the city centre; it did not extend as far as York Road station. A flat fare of 5p was charged for any journey on the route. By 1978 it had been decided to extend the service out to York Road station.

'Citylink' was essentially a Citybus service, but in 1985 it was taken over by NIR and was renamed 'Rail-Link'. The marketing of the service was taken over by NIR and the service promoted as the 'Belfast Inter-station – City Centre Bus Service'. The buses used on it were turned out in a version of the then NIR 'Intercity' silver and blue livery, with the NIR logo and the words 'Rail-Link' on the sides. A separate fare was still charged to rail travellers using the service,

though discounts were available to weekly or monthly ticket holders.

There was another change in 1988. NIR was 'sectorised', British Rail style, and the 'Rail-Link' service came under the control of NIR's new 'Suburban' sector and new buses introduced (see above). These were fitted with two-way radios to allow constant contact with the stations, so that they could better co-ordinate their connections with the trains. The biggest change was that the service was now free to rail users, the cost of using the bus being included in the price of the rail ticket. The buses were also fitted, during 1988, with the new railway computerised PORTIS ticket system machines, so that a ticket for a complete rail journey could be bought on the bus.

Although leased from Citybus, maintained by

Citybus, driven by Citybus drivers and using the same basic type of bus as Citybus, these were railway-operated buses in every important sense. They were under the day-to-day operational control of NIR and they were fitted out to the same standards, and in the same style, as the NIR suburban rail stock. They could not be used by Citybus for any other work without NIR approval.

The Rail-Link bus timetable appeared at the back of the railway timetable and, until the opening of the cross-river rail link at the end of 1994, the buses were scheduled to meet the trains at both Central and York Road/Yorkgate stations, Monday to Saturday, between 08.00 and 20.00.

Major railway capital works were completed in November 1994, linking Yorkgate halt (a new station which replaced the nearby York Road terminus) with Central station. Great Victoria Street station, in the heart of Belfast, was re-opened in September 1995. When the cross-river link between Yorkgate and Central opened, the four Rail-Link buses were re-deployed to operate a more frequent service between Central station and Belfast city centre. This service was retained in spite of the re-opening of the more centrally located Great Victoria Street station.

In the early days of the Citylink service all kinds of 'weird and wonderful' vehicles could be found operating it. At that time, because of heavy losses of buses due to terrorist action, Citybus strongly resembled the Lough Swilly, in having a splendid variety of second-hand cross-channel vehicles in service. Buses which operated the original Citylink service, for example, included ex-Edinburgh Corporation Tiger Cubs, ex-Ribble Leyland Leopards, ex-London Transport AEC Merlins, as well as almost life-expired ex-Belfast Corporation and ex-Barrow-in-Furness Daimler Roadliners.

Once the Rail-Link service, under NIR auspices, was set up to replace the Citylink one, it was operated by a number of standard Citybus Bristol RELLs, dedicated to the service and painted, as noted above, in the NIR 'Intercity'-style livery. Interior furbishment remained of the standard, and at that time spartan, Citybus variety.

When the service later came under the wing of the NIR Suburban sector, new buses were introduced. These were Leyland Lynxes and Tigers, with Alexander N-type bodywork, similar to buses introduced at the same time into the Citybus fleet. They were painted in the then new Suburban sector livery of red, orange and cream, with the NIR logo with the words 'Rail-Link' on the front and the words 'Suburban' and 'Rail-Link' with the NIR logo on the side. Interior furbishment was based on that of the trains and was to a very high standard, far superior to anything at that time in Citybus service, even to the extent of having piped music laid on!

In 1993 they were replaced by Leyland Cityliners. These again were similar to the new Citybus vehicles, with Alexander Q-type bodywork, and to the same high standard of interior and exterior finish as the Lynxes and Tigers. The livery used was similar to the Suburban livery, although NIR was no longer sectorised and the Suburban livery had been removed from the trains. The NIR corporate livery was by now a variation of the former 'Intercity' sector blue and silver, with black and yellow lining. Interestingly, the Rail-Link buses appeared in that livery in some of the NIR timetable artwork, but never on the vehicles themselves. All Rail-Link vehicles carried Citybus fleet numbers only.

In addition to the Rail-Link vehicles, in 1989 NIR Intercity sector purchased five 1984-built Van Hool-Alizee bodied 53-seater Leyland Tiger coaches from England. These had been operated originally by Smiths Shearings of Wigan. They were a high-floor design, with air suspension.

These coaches were bought for use as emergency rail replacement services, a particular problem during the years when the Belfast-Dublin line was frequently disrupted by terrorist explosions and bomb scares. The coaches were based at Craigavon Ulsterbus depot and were operated and maintained by Ulsterbus on contract. They carried NIR fleet Nos 1 to 5, but were also allocated Ulsterbus fleet Nos 591-5, used for internal Ulsterbus purposes.

They were turned out in a version of the NIR Intercity sector livery. The bus version of this livery consisted of a silver body, with two broad parallel flashes, one yellow, one blue, applied right around the vehicle. The NIR logo and the word 'Intercity' appeared on the side at waist height, just behind the cab. The NIR logo and the NIR fleet number appeared on the front of the vehicles.

NIR disposed of these vehicles to Ulsterbus in 1991, when it was decided that there was no longer a

Translink Volvo B10 in Centrelink livery in Donegall Place, Belfast.   Author

need for them, as Ulsterbus had agreed to provide buses, as and when required, for emergency work.

Translink

In 1995 Ulsterbus, Citybus and Northern Ireland Railways were brought under a common board of management. The brand name, 'Translink' was adopted for the new integrated public transport service to be offered by these co-ordinated companies.

The superficial similarities between Translink and the UTA are quite striking. Although the bus and railway companies at present maintain their separate identities, they are integrated at a senior management level. Many of the top management posts are now held by people from the road transport side, and a number of them have responsibility for aspects of both road and rail provision. For example, Ulsterbus Area Managers are responsible for aspects of both road and rail services (if any) in their areas.

As a result, Translink's new structure is converting the once largely independent companies into the operating divisions of a single integrated entity.

In 1997 NIR's Belfast 'Rail-Link' service was replaced by a new Translink Citybus service, in concept not unlike the original 'Citylink' service of 1976. This service, branded 'Centrelink', is sponsored by the DoE Transport Policy Branch.

The new Volvo buses dedicated to this service are turned out, not in the NIR livery of the old 'Rail-Link' service, but in an all-over 'Marks and Spencer' green livery. The 'Centrelink' service links the Europa Bus Centre/Great Victoria Street railway station with Central station, the new Laganside Bus Centre and Belfast city centre.

A similar road-rail link service in Derry connects Waterside railway station and Foyle Street bus station.

At the time of writing, Translink is beginning to integrate its road and rail activities by the provision of further bus services linking with rail services ('Linkline' services), and publishing joint road-rail timetables.

It has also promised integrated ticketing and the development of additional integrated road-rail termini. Translink claims that it has begun to eliminate wasteful competition between road and rail services. It has introduced a new joint road-rail parcel service, being marketed under the brand name 'Parcel-link'.

Bus and rail operations are increasingly marketed together, under the new banner, and Translink is having some success in establishing the new name as the recognised symbol of public transport in Northern Ireland.

The bus fleet now operated by Translink has been integrated between Ulsterbus and Citybus for many years. Most bus types, sometimes with minor differences, are common to both fleets. Apart from a handful of double-deckers used on tours work, the whole fleet consists of single-decker vehicles.

The stage carriage fleet comprises the following main types of buses:

- Bristol RELLs with Gardner engines and Alexander's bodywork.
- Leyland Leopards with Alexander's bodywork. These vehicles are only used by Ulsterbus.
- Leyland Tigers with Alexander's N-type bodywork.
- Leyland Tiger Cityliners with Volvo engines and Alexander's Q-type bodywork, in both bus and coach versions.
- Volvo B10Ms with Alexander's Q-type bodywork
- Dennis Darts with Wright's 'Easybus' and 'Handibus' bodies.
- Volvo B10Ls 'Ultra' low floor with Alexander's or Wright's bodywork.
- Volvo B10M articulated buses with Van-Hool bodies.
- Mercedes 0405s with Mercedes bodywork. There are currently five standard 43-seat single-deckers and four 59-seat articulated buses in service.
- A range of Wright's bodied Mercedes minibuses, operated by the Ulsterbus private hire subsidiary Flexibus and by Ulsterbus on 'Busybus' town service work.

An attractive new livery is now being applied to some of the Translink bus fleet. At the time of writing (March 2000) only some of the low-floor Volvos are turned out in it. It is a livery very different from the traditional Citybus colours of red and cream, which has its origins in tramway days, or the Ulsterbus blue and ivory which dates back to the company's foundation in 1967. Like the BOC, NIRTB and UTA liveries, it is basically two-tone green (see page 128).

This new livery, as applied to the Volvos, consists of a mint green applied to the front, sides, and roof of the cab area, with most of the rest of the bus finished in the darker 'Translink' green. The mint green cab is divided from the 'Translink' green area by a thick red band running vertically up the sides and across the roof. A red flash, curving from just behind the front wheel to the lower rear of the bus, is applied to both sides of the vehicle. This red flash is mirrored by a mint green one starting from the same point, but which curves up to the roof edge. The upper back pillars are finished in red, as is the front bumper section. The upper back of the bus is finished in mint green, and the lower back panels in 'Translink' green.

The word 'Citybus', with the Translink 'T' symbol, is applied above the first window behind the cab, while the slogan 'a Translink service' is applied on each side above the rear windows. Lettering is white. Fleet numbers are applied to the front and rear, using black numerals.

In the Ulsterbus version of this livery, the red features are replaced by blue (see page 128).

Because of the low seating capacity of the new low-floor single-deckers, the economics of double-deckers are once again looking attractive for city work. Recently Translink has experimented with a double-decker in Belfast on a trial basis. In December 1999 a Dennis Trident with Alexander's bodywork was assessed. At the time of writing, it is not clear whether this will lead to any orders.

Chapter 7

Surviving vehicles

Steam railmotors

A number of the vehicles or types of vehicles mentioned in this book have survived into preservation. The carriage portion of former BCDR steam railmotor No 2 has been preserved by the Downpatrick Railway Museum project. It has been mounted on a suitable length GSWR carriage chassis, the railmotor's original chassis having been cut up for scrap in 1956. The carriage body had survived as a residence until 1986, when it was acquired by the Downpatrick project. It is due to be restored to the form in which it finally ran, as open brake third No 72.

Half of the body of one of the GNR auto-train trailers, No 15 or 16, used on the Dublin-Howth and Belfast-Lisburn services, has survived as a privately owned summer house at Gyle's Quay, near Dundalk.

Railcars

A number of narrow gauge and standard gauge railcars have survived, particluarly examples from the County Donegal Railway. Most of the narrow gauge vehicles are located in the Ulster Folk and Transport Museum and in the Foyle Valley Railway centre in Derry. The few remaining standard gauge vehicles can be found at various locations around the country.

Clogher Valley Railway Walker's

Above: Coach portion of BCDR railmotor No 2 at Downpatrick Railway Museum.

Author

Right: Half of body of GNR auto-train trailer, either No 15 or 16, at Gyles Quay near Dundalk.

Author

railcar No 1 (preserved as CDR No 10), the pioneer CDR Allday and Onions railcar (No 1) and CDR railcar trailer No 3 (originally DBST) are part of the Ulster Folk and Transport Museum's collection at their excellent rail gallery at Cultra.

NIR BRELL/British Leyland Railbus No R3 is currently (2000) on loan to the UFTM and is on display at their Cultra Rail Gallery. There are plans to move it to Downpatrick Railway Museum during 2000 or 2001. (BRELL/British Leyland railbus No R2 is located on a private 4' 8½" railway at Riverside Mill, between Dundalk and Greenore.)

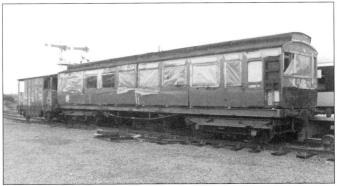

CDR Walker articulated railcars Nos 12 and 18 are in the Foyle Valley Railway collection and operate on the line being laid towards St Johnston from the FVR station at Craigavon Bridge, along the track bed of the GNR 'Derry Road'. Both railcars have been extensively restored since coming into the ownership of the FVR.

Other survivors

The South Donegal Railway Restoration Project, based in Donegal town in the Republic of Ireland, has acquired CDR railcar trailer No 5, which it is in the process of restoring.

CDR Walker articulated railcars Nos 19 and 20 operate on the Isle of Man Railway. They were bought by the IoMR in 1961, when the CDR ceased its railway operations.

The coach portion of West Clare railcar No 3386 is operated by Bord na Mona on one of its tourist trains at the Bellacorick Bog Railway, in Co Mayo. A West Clare railcar bus-coach trailer is operated by the Cavan and Leitrim Railway preservation group at Dromod.

LMS (NCC) railcar No 1 is under restoration by the Railway Preservation Society of Ireland at Whitehead.

SLNCR railcar B is in store at the Great Southern Railway Preservation Society site at Mallow, in the Republic of Ireland. At the time of writing no restoration work has taken place to this vehicle.

Railbuses

GNR railbus No 1 is part of the UFTM collection in Cultra. This railbus was extensively restored by Grimley Brothers of Annaghmore. After its restoration, it ran under its own power on 21 March 1993 from Poyntzpass, on the ex-GNR main line, to the new Rail Gallery at Cultra, where it is now on display.

GNR buses

A number of the types of road bus mentioned in the text have survived into preservation.

At least five GNR single-deck buses have survived. All had been taken into CIÉ stock in 1958. One, No 324, a GNR-Gardner, built in 1941 and rebodied in 1949, is part of the UFTM collection, but is not currently on display. A second, No 390, a GNR-Gardner with a Park Royal/GNR body, is on display at the TMSI museum in Howth, as is No 274, an AEC Regent IV, built in 1954. Two others, Nos 389 (currently at Dromod, Co Leitrim) and 397 are privately preserved. No 427, an AEC Regal III coach and No 438 GNR AEC Regent III 56-seater double-decker, with a Park Royal/GNR body, are also in the TMSI collection in Howth.

All the above buses are preserved in the GNR Oxford blue and cream livery.

NCC buses

LMS (NCC)-built Leyland TS6 No 14, one of a batch of 18 similar vehicles supplied to the NCC in 1934, later NIRTB No M402/M714, is privately owned by Ken Alexander. This bus was originally fitted with a 32-seater Weymann bus body, when in railway service, but was rebodied by the NIRTB with its standard 34-seater body in 1944. It is preserved in NIRTB livery.

NIRTB and UTA buses

A variety of NIRTB and UTA-built buses survive, mostly in private ownership. Five are Leyland Tiger PS1 single-deckers, built by the NIRTB in 1947. No A515, owned by Ulsterbus, is now preserved in UTA livery, as are Nos A8517 (owned by James Begley of Lisburn) and A8556 (owned by John Magill of Eglinton, Co Londonderry). No A8560, privately owned by WH Montgomery of Belfast, carries its original NIRTB

livery. No A8570 belongs to the Transport Museum Society of Ireland and is in store at their Castleruddery site in Co Wicklow.

A 1949 UTA-built Leyland PS2/1, No C8858, is privately owned by Thomas Montgomery of Belfast. This bus was originally fitted with a standard 34-seat single-deck body but, about 1965, was modified to a front entrance layout, with the seating increased from 34 to 37, for Derry City service work. It survived into Ulsterbus ownership, for use by the driving school, and was the only PS2 to be painted in Ulsterbus colours. It is now preserved in UTA light green livery.

A UTA 1956 Leyland PSUC1/5T Tiger Cub single-decker bus, No K318, now painted in UTA dark green livery, is privately preserved by the Bell family. Another UTA Tiger Cub, No H301, built in 1954 with Saunders Roe bodywork, originally a Leyland demonstrator, is privately owned by Bryan Boyle of Belfast and Jim Poots of Portadown.

Leyland Titan PD2/1 double-decker No D927, built in 1950, preserved in UTA livery, is privately owned by Irwin Miller of Ballyclare. PD2/10 No M659, restored to UTA livery, is owned by Thomas Mahood of Portavogie. This bus was re-bodied as a double-decker in 1958, by the UTA, on the chassis of semi-decker No B8940, built in 1948. It is fitted with the same type of 60-seater body as was fitted to the PS2s rebuilt as double-deckers.

A UTA PD3/4 double-decker, No R815, built in 1962, has been restored to its UTA light green livery and is owned by Raymond Begley of Lisburn.

Ulsterbus has preserved three vehicles. As well as A515 mentioned above, the company also owns a 1942 Bedford OWB Utility-bodied single-decker bus No V957. This bus is in the livery of the NIRTB, to whom it was originally supplied, but the body is totally new, having been rebuilt by Ulsterbus workshops for the 50th anniversary of the NIRTB in 1985. The third Ulsterbus vehicle is a UTA AEC Reliance single-decker coach, No S234 of 1963, which survived into Ulsterbus use. It has been restored to the two-tone blue livery applied to the UTA 'semi-luxury' coaches, when built in the 1960s.

LLSR buses

No 75, a LLSR Royal Tiger single-decker, with Saunders Roe centre entrance body, had been

Top: Standard NIRTB/UTA Leyland PSI single-decker design in NIRTB livery at the ITT's Bangor rally in 1999. Author

Centre: Leyland PD3/4 No R815 at a vintage commercial vehicle rally near Downpatrick Railway Museum. This is the only survivor of this once numerous type.

Author

Bottom: NIRTB Bedford OWB, rebuilt by Ulsterbus in 1985 to mark the 50th anniversary of the foundation of the NIRTB.

Author

preserved but, unfortunately, it was recently destroyed by fire in Galway. It had carried the LLSR maroon and silver livery.

CDR buses

Although no CDR buses survived into preservation, at least two CIÉ single-deckers, similar to those used by the CDR, have survived. P347, a Leyland OPS2/14 with CIÉ bodywork, built in 1953, is owned by the TMSI and is on display at their Howth museum; and E14, a Leyland L2 Leopard with CIÉ bodywork, built in 1961, is preserved by CIÉ at their Broadstone depot.

The UTA Saro-bodied Tiger-cub No H301, preserved by Bryan Boyle, as noted opposite in the UTA section, is similar to the CDR Tiger-cubs bought from England to replace the hired CIÉ P-classes.

CIÉ

In addition to the ex-GNR buses already mentioned, a number of CIÉ vehicles have been preserved by the Transport Museum Society of Ireland either at their Howth or Castleruddery sites. These include the following:

Year new	Make and model	Original owner	Fleet No	Converted to:
Single-deckers				
1933	Dennis Lancet 1	DUTC	F21	
1953	Leyland Tiger PS2	CIÉ	P347	
1954	Leyland Royal Tiger PSU1 (coach)	CIÉ	U10	
1954	Leyland Royal Tiger PSU1	CIÉ	U78	
1964	Leopard L2	CIÉ	E170	
1965	Leopard PSU3	CIÉ	C17	
1966	Leopard PSU4	CIÉ	C231	
1968	Bedford VAS5	CIÉ	SS1	
Double-deckers				
1937	Leyland Titan TD4	DUTC	R1	
1948	Leyland Titan PD2/3	CIÉ	R389	
1953	Leyland Titan OPD2	CIÉ	R506	
1955	Leyland Titan OPD2	CIÉ	R567	
1960	Leyland Titan PD3	CIÉ	RA105	
1961	AEC Regent V	CIÉ	AA2	
1964	Leyland Titan PD3	CIÉ	R911	
1965	Leyland Titan PD3	CIÉ	R920	
1966	Leyland Atlantean PDR1	CIÉ	D44	
1975	Leyland Atlantean AN68	CIÉ	D694	
Service vehicles (former buses)				
1930	Leyland Lion LT2	GNR	271	GNR lorry No 150
1951	GNR Gardner	GNR	G387	Ambulance bus
1953	Leyland OPS3	CIÉ	P193	Tow wagon
1958	Leyland OPD2	CIÉ	R819	Tow wagon

Chapter 8

Towards the twenty-first century

In previous chapters, it has been shown how the balance between road and rail has changed over the years, and how the railway companies have sought to respond to the changing circumstances in which they found themselves having to operate. An important theme of this work, of course, is how these developments were reflected in the changing mix of vehicles used, changing vehicle design and development, and the use of new types of passenger carrying vehicles.

It is clear to observers of the transport scene that these processes of change are still ongoing, and the patterns of passenger-carrying provision, which will be in place in the early years of the twenty-first century, are now starting to emerge. In trying to identify the likely detail of this provision, it is important to remember that the fashions and trends in passenger-carrying systems will not be set in Ireland, but in mainland Europe, the USA and the Far East. As always, we in Ireland will be followers of fashion, rather than trend-setters.

As we have seen, the methods of passenger-carriage in the past, as in the present, were partly the result of changes in technology and partly a matter of fashion and trend which were in turn only indirectly related to this technological change.

Initially, the technologically revolutionary railways were all-powerful. Road transport was a very poor substitute, which posed no serious threat to the railways' position until after World War I. Motor cars were non-existent until the latter years of the nineteenth century, and the poor roads inhibited the development of the early omnibus services to any degree, outside the main towns and cities. Even in urban areas, tramway systems were the preferred form of mass transit.

Railways in this period were anxious to attract all possible traffic to rail. In situations where it was uneconomic to construct and operate a conventional heavy rail system, they promoted light railways or tramways.

Railways were seen, by national and local decision-makers, as modern. They were the bringers of prosperity and civilisation and the mark of a community belonging to the national or international mainstream. Railways were high fashion and the promotion and construction of railways was big business, glamorous and always in the news.

After World War I, this pattern began to change radically. Again, the root of the change was technological. It was the result of the development of the internal combustion engined vehicle, during the war, coupled with dramatic improvements in road-building technology. There was also a growing acceptance that the provision and upkeep of roads, though not the railways, should be a function and responsibility of the state, at county and later at national levels, rather than of the parish or toll road company.

As a result, the railway companies found themselves under growing pressure from an increasing number of road competitors, in an emerging industry where 'entry' costs were low and where statutory regulations on safety and conditions of employment were minimal. In addition, whereas railways had to bear their full track costs, their road competitors did not.

The railways responded, as we have seen, in two main ways, by developing road services of their own and by developing more cost-effective railway vehicles, ultimately based on the use of the internal combustion engine.

We have also also seen, during this period, the state being drawn increasingly into the picture, as the railways struggled to survive the growing threat from cheap and flexible road transport rivals. The modern preoccupation of the state with the 'correct' balance between road and rail provision, and whether or not the state should subsidise public transport services for social or economic reasons, began during these inter-war years.

The third stage, from which we are now beginning

to emerge, saw the balance of advantage shift decisively to road transport and the railway network being reduced drastically.

Vast sums of public money were ploughed into improving the existing road network and creating a new network of motorways, while railways were closed, often to save relatively small amounts of taxpayers' money.

Roads, particularly motorways and urban ring roads, were now the fashion and even Northern Ireland, with only 1½ million of a population, planned a system of seven motorways radiating out of Belfast, with an associated elevated urban ring road, built to motorway standard.

There was (and is) an implied assumption that everyone had the right to own and use private transport to whatever extent they could afford, and it was the job of the state to provide the road capacity to meet this demand and to re-organise the urban landscape to facilitate the use of the motor vehicle.

Public transport systems were regarded as second-class systems and were given a lower priority for any investment funds available. As a result, in the second half of the twentieth century they suffered a slow but steady decline in quality and value for money. City tramways, in particular, were closed in the 1940s and 1950s because they 'interfered' with the 'free movement' of private cars. The same fate befell Belfast's trolleybuses in the late 1960s.

Railways in Ireland have survived only by being taken into state ownership, often as part of an overall state transport organisation. In Northern Ireland, railways very definitely took second place to road transport provision. In one famous remark, Mr William Craig, the Minister for Home Affairs in the Northern Ireland government of the 1960s, at whose behest the Benson Report, mentioned earlier, was produced, stated that the railway, as a mode of transport, was "as obsolete as the stagecoach". He was not alone in this view in decision-making circles.

Railway motive power technology saw the abandonment of steam and became firmly based on the internal combustion engine and electric traction which, as we have seen, had been pioneered by the independent railway companies, mainly during the 1930s.

From the late 1960s the organisational fashion changed from the public sector, public corporation model, offering an integrated road/rail provision and

subsidised where necessary from the public purse, to one based on a private sector, limited company, competition-based and profit-driven business model. Road and rail provision were separated and placed under the control of what were semi-private companies, operating within a purely commercial framework and competing with each other, as well as with outside providers.

At present, in 2000, another sea-change seems to be under way. The very success of road transport, in particular of the motor car, has created problems for society as a whole in the form of noise, pollution, traffic congestion and high levels of road-accident related injuries and deaths.

Increasingly, people are beginning to question the whole emphasis on a private car/heavy lorry-dominated transport system. They are becoming uneasy with the uncontrolled growth of road transport, with its implications of yet more road-building and widening programmes which disrupt communities, damage landscapes and which have to be funded by the public purse.

There has long been a belief, now recently backed by official evidence, that all road building achieves is to create space for ever more cars and trucks, without actually solving the problems of traffic congestion. As a result, the pendulum of public and political opinion seems to be now starting to swing the other way again, towards a greater emphasis on public transport provision, as well as, and perhaps eventually in preference to, private transport.

This gradual, but accelerating shift in emphasis from private to public transport is being reflected in a growing opposition to major new road-building programmes, particularly those involving motorways or those disrupting urban areas, or damaging areas of natural beauty. There is a growing support amongst the public for good quality public transport provision.

One beneficiary already of this change has been the revival of Light Rail or Light Rapid Transit (LRT) in Britain and the USA, where tramways had largely disappeared in favour of road-based transportation systems.

In the United Kingdom and Ireland, systems half-way between light and normal heavy rail have been built in Newcastle-on Tyne, London Docklands and Dublin. In addition, the tram has been reborn in the form of LRT in Manchester, Croydon and Sheffield and, as funds and government approval become available, they are likely to make a comeback in a number of other British cities. Half-a-century after their demise, trams are once again back in fashion. Indeed, in a BBC radio interview in March 2000, the chairman of the CBI called for the establishment of more LRT systems in British cities.

An LRT system, 'Luas' (the Irish word for speed), is planned for Dublin and the first phase of this new network is due for completion in the early 2000s. A proposal, on paper, for an LRT system for Belfast also exists, although it is unlikely to be built in the near future, if ever.

In Northern Ireland, the railways have come back from the brink of extinction, with increased investment in permanent way and rolling stock and, in Ireland as a whole, investment in railways has been on the increase, supported by significant funding from the European Union.

In Northern Ireland, this change of attitude has been evidenced by the building of the cross-harbour rail link, which opened in late 1994, and the rebuilding of Great Victoria Street station in central Belfast, which opened in 1995.

European Union money has also been made available to upgrade the Belfast-Dublin rail link, which has been virtually rebuilt. New, powerful, General Motors locomotives and modern, French-built, rolling stock have been purchased for it in a joint venture between Iarnród Éireann and Northern Ireland Railways. In the summer of 1999 the Belfast Central route was completely rebuilt. The re-building of the Bleach Green to Antrim section of the ex-NCC main line is underway, and the Belfast to Bangor line is to be completely re-laid.

In the Irish Republic major investment is going into significantly improving the country's rail system. In 1998 £63 million was allocated to expand the DART system and £26 million additional investment was allocated to mainline rail.

In January 1995, the Northern Ireland Office announced the abandonment of a major road scheme through the Belvoir Forest beauty spot in south Belfast. This proposal had attracted fierce opposition from many quarters and its abandonment was regarded as a major victory for the local anti-roads lobby. The announcement was made even more significant by including a statement that the province's two major bus companies, Ulsterbus Ltd and Citybus Ltd, were to be merged with Northern Ireland

Railways Ltd to create again, after a gap of 30 years, an integrated public transport service to be called 'Translink'. It was also announced that, in future, an increased emphasis would be placed on the development of public transport, and a reduced emphasis on the provision of new roads.

In January 1995 NIR announced their intention to replace their '80-class' diesel-electric units, purchased in the mid 1970s, with diesel multiple units. These will either be similar to the Iarnród Éireann 'Arrow' trains, or based on the British Rail 'Sprinter' sets. In 2000 these still have not been ordered but are likely to be funded under the government's Private Finance Initiative or PFI.

All new railway structures, such as bridges and workshops, are being built with sufficient overhead clearances to allow for the installation of overhead, electric wires.

In Dublin, outside the area served by the DART, system, suburban rail services have been improved. Stations long closed have been re-opened and are being served by a new generation of diesel-multiple units introduced in 1994 by Iarnród Éireann, consisting of a fleet of new 'Arrow' railcars built in Japan and Spain.

Other possible future patterns of transport provision may be discerned from developments taking place elsewhere in the world. Examples of these include the introduction of road pricing schemes, which will require road users to bear a larger proportion of track costs and will act as a deterrent to car use within urban centres.

In addition, banning traffic from city centres, banning road freight traffic at certain times, or from travelling more than a certain distance, has been introduced or is being considered in a number of places. Belfast already has the advantage of having most traffic banned from the city centre, although the original reason for this was security-related.

Other changes may include the imposition of a carbon tax on fuel as an anti-pollution measure, and proposals to change the tax structure, so that a larger proportion of the costs of operating private cars becomes variable with the amount they are used. This will change the balance in the proportion of fixed to variable costs and make the costs of car use clearer to members of the public. This should allow more realistic comparisons to be made with public transport alternatives.

Proposals to wholly, or partially, privatise public transport provision have been made, both in Northern Ireland and, more recently, in the Republic of Ireland. If this does happen, much could be learned from the mistakes made in British rail privatisation in the recent past.

For many of these ideas to work properly, improved public transport provision and higher standards of comfort will be required. Public transport providers, both north and south of the border, recognise this and, within the inevitable limits of public sector funding, are slowly raising standards and public expectations of public transport services. Evidence of this is to be seen in the introduction of higher quality buses and trains, bus priority systems such as bus lanes and QBCs and improved schedules for both road and rail services..

Private road transport will never return to the minor role that it played a century ago. However, the decline in public transport in general, and rail transport in particular, is ending. It is being re-discovered that, for certain types of traffic, the railway, as it always did, offers an economically and environmentally superior form of transportation to road based systems.

As a result, it is unlikely that any further rail lines in Ireland will be closed and it is clear that usage of those still in existence is on the increase. As we enter the new century, we are beginning to see serious proposals being made for the opening of new routes, light and heavy rail, and the re-opening of routes closed during the low period in the 1950s and 1960s. As has been noted, tentative proposals for such re-openings are appearing in the Irish Republic.

As things stand at present, part of the closed Harcourt Street-Bray line in Dublin will be re-opened as part of the Luas LRT system. There are also plans in Belfast to build the 'E-Way' guided bus way, along the Holywood Arches to Dundonald section of the long closed BCDR main line.

In Belfast, Translink is developing plans for a new 'super route' bus way to allow buses to by-pass traffic congestion in the Saintfield Road area of Belfast, and for a significant extension in the provision of bus lanes in Belfast In the twenty-first century, as in the twentieth, public transport managements will continue to evolve new and innovative solutions to the transport problems with which they are faced.

Artist's impression of the proposed Translink E-way vehicle, posed in front of Belfast City Hall.

Translink

Appendix
Railway company bus routes

LMS (NCC) routes in 1932

(Towns with LMS (NCC) railway stations or halts are shown in *italics*.)

Route
Nos

1	*Belfast* and Islandmagee (via *Carrickfergus* and *Whitehead*)
2	Browns Bay and *Larne* (via *Ballycarry*)
3	*Belfast* and *Ballycarry* (via *Monkstown* and *Belfast* and *Larne*, Wednesdays and Sundays only)
4	*Belfast* and Cushendun (via *Larne*)
4A	*Belfast*, *Ballycastle*, Giants Causeway and *Portrush* (summer only)
5	*Belfast* and *Larne* (via *Mossley*, Straid and Glenoe)
6	*Belfast* and *Larne* (via *Ballyclare*, *Ballynure* and Millbrook)
7	*Larne* and Halls Crossroads
8	*Belfast*, *Doagh* and *Ballyclare* (via Glengormley)
8A	*Belfast* and *Ballyclare* (via *Whitehouse* and *Mossley*)
9	*Belfast* and *Ballyclare* (via Straid and *Ballynure*)
10	*Doagh* and *Ballyboley* (via *Ballyclare* and *Ballynure*)
11	*Belfast*, *Ballymena*, Cushendun (via *Ballyclare*, *Doagh* and *Parkmore*)
11A	*Ballymena* and *Rathkenny* (via Quarrytown, Sundays only)
12	*Belfast* and *Ballymena* (via Glengormley and Antrim)
12A	*Ballymena* and Bunkers Hill (via Slatt, Sundays only)
13	*Belfast* and *Cookstown* (via *Antrim*, *Castledawson*, *Magherafelt* and *Moneymore*)
14	*Belfast* and *Cookstown* (via *Antrim*, *Staffordstown*, Ballyronan and Coagh)
15	*Kilrea* and *Antrim* (via Portglenone)
16	*Belfast* and *Maghera* (via *Antrim* and *Castledawson*)
17	*Belfast* and *Draperstown* (via *Antrim*, *Castledawson*, *Magherafelt* and Tobermore)
18	*Magherafelt* and *Draperstown* (via *Desertmartin*)
19	*Ballymena* and *Larne* (via *Moorfields* and Millbrook)
20	*Ballymena*, Carnlough, Glenarm (via Broughshane)
21	*Ballymena*, Portglenone, *Magherafelt*, *Randalstown*
22	Bellaghy and *Portrush* (via Portglenone, *Ballymoney* and *Coleraine*)
23	*Cookstown* and *Portrush* (via *Castledawson* and *Kilrea*, Sundays only)
24	*Ballymoney* and *Portrush* (via Bushmills, Sundays only)
25	*Coleraine* and *Ballycastle* (via *Dervock* and Liscolman)
26	*Coleraine* and *Portrush* (direct)
27	*Coleraine* and *Portstewart* (direct)
28	*Portrush* and *Portstewart* (direct)
29	*Coleraine* and *Kilrea* (via *Ballymoney* and Rasharkin)
30	*Coleraine* and *Magherafelt* (via Castleroe, *Kilrea* and *Maghera*)
31	*Coleraine* and *Kilrea* (via Macosquin and Landmore)

32	*Coleraine* and *Maghera* (via *Garvagh*)
33A	*Limavady* and *Dungiven* (via Terrydremond and Camnish)
33B	*Limavady* and *Dungiven* (via *Drumsurn* and Scriggan)
33C	*Limavady* and Myroe
34	*Londonderry* and Alla (via Ardmore and Ballyarton)
35A	*Londonderry* and *Dungiven* (via Claudy and Straidarran)
35B	*Londonderry* and Fenny (via Claudy and Park)
35C	*Londonderry* and *Dungiven* (via Claudy and Holly Hill)
36	*Belfast* and *Ballycastle* (via *Ballymena* and *Armoy*, Sundays only)
37	*Belfast, Ballymena* and *Portrush* (via *Antrim, Dunloy, Ballymoney* and *Coleraine*, Sundays only)
38	*Londonderry, Limavady* and *Portrush* (via *Eglinton, Castlerock* and *Coleraine*, Sundays only)

GNR bus routes (GNR bus routes had no route numbers)

(Towns which had railway stations at any time are shown in *italics*.)

Northern Ireland 1933-34

Belfast-Dundalk-Drogheda-Dublin
Belfast-Lisburn
Belfast-Lisburn-Maze-Greers Corner-Culcavey-
    Moira-Magheralin-*Lurgan*
Belfast-Lisburn-Newry
Belfast-Aghalee-*Lurgan*
Belfast-Lurgan-Portadown-Dungannon-
    Ballygawley-Omagh

Belfast-Lurgan-Gilford-*Tandragee*-Markethill
Belfast-Crumlin-Diamond
Armagh-Milford-Middletown
Armagh-Keady
Armagh-Castleblaney
Lurgan-Kilmore
Warrenpoint-Rostrevor-Kilkeel

Republic of Ireland post 1935

Dublin-Drogheda-Dundalk-Newry
Dublin-Slane-*Ardee-Carrickmacross*
Dublin-Ashbourne-*Navan-Kingscourt-*
    Carrickmacross
Dublin-Navan-Kells-Cavan-Enniskillen
Dublin-Howth
Dublin-Howth (via Dollymount)
Dublin-Sutton Cross-Strand Road
Dublin-Portmarnock-Malahide
Kinsealey-Baldoyle (church services)
Dublin-Hole in the Wall
Skerries Station and Town
Dublin-Skerries (via Swords, *Rush and Lusk*)
Donabate Station-Portane
Portrane-Swords
Rush and Lusk Station-Rush (town)
Skerries-Loughshinney-Rush (church services)
Drogheda-Baltray-Clogher-Togher

Drogheda-Tenure
Drogheda-Collon-*Ardee*
Drogheda-Bettystown-*Laytown*
Drogheda-Balbriggan-*Drogheda*
Drogheda-Slane-Sallygardens Cross-Woodtown
    Cross
Drogheda-Dromin
Drogheda-Stamullen-Bellewstown
Drogheda Station-Drogheda town
Drogheda-Ardee (via and *Dunleer*)
Drogheda-Duleek-Kentstown
Drogheda-Slane-Gormanlough
Drogheda-Drumconrath-*Ardee*
Dundalk-Stabannon-*Ardee*
Ardee-Blackrock (via Knockbridge)
Drogheda-Ardcath-Garristown
Drogheda-Dowth
Ardee-Aclint Cross

Drogheda-Blackhall Cross
Drogheda-Ardcath-Kilmoon Cross
Drogheda (John St)-Donore-Rossnaree Cross
Ardee-Churchtown Cross
Ardee-Duffy's Cross-*Dundalk*
Drogheda-Grangebellew
Dundalk-Ardee (via Louth)
Dundalk Town Services
Dundalk-Blackrock-Dromiskin-*Dunleer-Ardee*
Dundalk-Castlebellingham-Annagassan
Dromin-*Dundalk*
Dundalk-Drumbilla-border
Dundalk-Corcreaghy
Monaghan-Castleblayney (via Cremartin)
Monaghan-Clogher border (via Tedavnet)
Ballybay-Shercock
Dundalk-Carrickmacross-Castleblayney-
    Monaghan
Dundalk-McShanes Cross-*Castleblayney*
Carrickmacross-Inniskeen
Cootehill-Blackrock (Co Louth)
Dundalk-Kilcurry
Dundalk-Carrickmacross-Bailieborough-*Cavan*
Greenore-Grange
Greenore-Boherboy
Dundalk-Greenore-Carlingford-Omeath-Newry
Ferryhill-*Dundalk* (via Ardaghy)
Dundalk-Corcreaghy-*Carrickmacross*)

Carrickmacross-Culloville
Carrickmacross-Ballybay
Cootehill-Ballybay
Dundalk-Dromiskin-*Castlebellingham*
Castlebellingham-Keady
Monaghan-Ballybay (via Ardgahy)
Monaghan-Tyholland-border
Virginia Road Station-Kilnaleck-*Cavan*
Cootehill-Monaghan
Cavan-Cootehill
Monaghan-Emyvale-Aughnacloy
Monaghan-Newbliss (via Greenans Cross)
Monaghan-Smithborough-Clones
Londonderry-Stranorlar-Donegal-Bundoran-
    Sligo
Ballyshannon-Bundoran
Ballyshannon-Lough Derg, mainland shore
Ballyshannon Mkt Y -*Ballyshannon GN station*
Donegal-Portnoo
Carrick on Shannon-Drumshambo-Bundoran
Sligo-Ballintrillick
Ballyshannon-Rossnowlagh
Killybegs-Glencolumbkille-Malinamore
Pettigo Station-Lough Derg, mainland shore
Ballybofey-*Letterkenny*
Stranorlar-Glenties-Portnoo
Glenties-Dungloe
Ardara-Portnoo (direct)

CDR bus routes in 1958

(Towns which had railway stations in 1958 are shown in *italics*.)

Stranorlar-Glenties-Portnoo
Ballybofey-*Letterkenny*
Malinmore-*Killybegs*
Glenties-Dungloe

CDR bus routes in 1971

(Towns which originally had railway stations shown in *talics*.)

Killybegs-Donegal-Belfast (via *Strabane* and *Derry*)
Letterkenny-Belfast (via *Strabane*)
Strabane-Ballybofey-Donegal-Killybegs
Strabane-Raphoe-Convoy-Letterkenny
Portnoo-*Glenties-Belfast*
Ballybofey-Glenties-Portnoo
Strabane-Donegal-Ballyshannon (for Bundoran and Sligo)
Donegal-Killybegs-Portnoo
Ballybofey-Letterkenny
Killybegs-Malinmore
Sligo-Donegal-Ballybofey-Derry
Ballyshannon-Rossnowlagh

Express Services operated jointly with Ulsterbus:
Letterkenny-Aldergrove Airport (Belfast)
Killybegs-Donegal-Ballybofey-Belfast (via *Strabane, Omagh* and *Dungannon*)
Dungloe-*Glenties-Donegal-Belfast* (via *Pettigo* and *Omagh*)

LLSR bus routes in 1953

(Towns which originally had railway stations shown in *italics*.)

Derry-Fahan-Buncrana
Buncrana-Ballyliffin-Carndonagh
Derry-Carndonagh-Malin Head
Derry-Moville-Shrove
Moville-Culdaff-*Carndonagh*
Derry-Letterkenny
Church Hill-Letterkenny

Derry-Letterkenny-Rosapenna-Downings
Derry-Letterkenny-Dunfanaghy-Gweedore-Dungloe-Burtonport (via Kildrum)
Dungloe-Bunbeg-*Falcarragh* via Foreland route
Fanad Head-Milford (via Rosnakill)
Glenvar-Rathmullan-*Letterkenny*

SLNCR bus routes in 1954

(Towns which had SLNCR railway stations shown in *italics*.)

Blacklion and *Sligo* (via *Manorhamilton* and Glencar)
Blacklion and *Sligo* (via Dowra, Drumkeeran and *Dromhair*)
Manorhamilton and *Dromhair* (via Cloonaquinn and Bohy, Saturdays only)
Dowra and Ballinaglera (Tuesdays, Thursdays and Fridays only)
Shanvas Cross (Manorhamilton) and *Sligo* (via Moragh, Fridays only)

Bibliography

Primary Sources
CIÉ Annual Reports and Accounts
Railway and bus companies' timetables
Urwick, Orr and Partners, Survey of Road Motor Services of the GNR
UTA Annual Reports and Accounts

Articles
Carse, SJ, 'Motive Power of The CDR-2', *Journal of the Irish Railway Record Society*, Vol 16
Abbott, James, 'Class 141 Railbus, Part 1', *Modern Railways*, Vol 40, No 412, January 1983
_____ 'Class 141 Railbus, Part 2', *ibid*, Vol 40, No 413, February 1983
Clements, R N, 'Steam Carriages of Inchicore', *ibid*, Vol 8
Coleford, IC, 'Sentinel Railcars (Part One)', *British Railways Illustrated*, Vol 3 No 10
Cuffe, P, 'CIÉ's Railcar Fleet-1 Mechanical', *Journal of the Irish Railway Record Society*, Vol 10
Davey, P, 'Railway Remnants Around Belfast', *ibid*, Vol 19
Gamble, NR, 'NIR Railcars: 2 – The MPD and Class 80 cars', *ibid*, Vol 14
Gould, MH, 'The Magilligan Branch', *Five Foot Three*, Winter '79/'80
Hibbs, John, 'Road Passenger Transport in Ulster', *Transport History*, Vol 3, No 2
Houston, JH, 'NCC *Locomotives* 1903-1933', *Journal of the Irish Railway Record Society*, Vol 3
Hunter, R, (letter), 'NCC Sentinel Railcar', *Five Foot Three*, Winter '94/'95
Kelly, Edward, 'From Road to Rail', *Journal of the Irish Railway Record Society*, Vol 4
Ludgate, RC, 'Fifty Years of Ulsterbus', *Buses*, Vol 38, No 370
Mears, John, 'New Iarnród Éireann railcars', *Journal of the Irish Railway Record Society*, Vol 18
Newham, AT, 'The Portstewart Tramway', *ibid*, Vol 6
O'Meara, J, 'The GNR Crisis', 1938, *ibid*, Vol 17
Stafford, JW, 'Development of Multiple-Unit Diesel Trains In NI', *ibid*, Vol 7

Books
Abbott, Rowland AS, *Vertical Boiler Locomotives and Railmotors*, Oakwood Press, 1989
Boyle, BC, *Buses in Ulster, Volume 1, The NIRTB*, Colourpoint Books, 1999
Carroll, Joe, *Through the Hills of Donegal*, South Donegal Railway Restoration Soc, 1992
Casserley, HC, *Outline of Irish Railway History*, David and Charles, 1974
Coakham, Desmond, *Belfast and Co Down Railway*, Midland Publishing, 1998
Corcoran and Manahan, *Winged Wheel*, Transport Museum Publications, 1996
Cummings, John, *Railway Motor Buses and Bus Services 1902-1933*, Oxford Publishing Co, 1978
Currie,JRL, *The Northern Counties Railway, Volume 1*, David and Charles, 1974
_____ *ibid*, *Volume 2*, David and Charles, 1974
Doyle, Oliver and Hirsch, Stephen, *Locomotives and Rolling Stock of Córas Iompair Éireann and Northern Ireland Railways*, Signal Press, 1979
Dougherty, Hugh, *The Bus Services of the Co Donegal Railways*, Transport Research, 1973
Ferris, Tom, *Irish Railways in Colour*, Midland Publishing, 1992
_____ *The Irish Narrow Gauge Volume 1*, Blackstaff Press, 1993
_____ *ibid*, *Volume 2*, Blackstaff Press, 1993
Ferris and Flanagan, *The Cavan and Leitrim Railway*, Midland Publishing, 1997
Flanders, Steve, *The County Donegal Railway* , Midland Publishing, 1996
Fleet History PI 2, Córas Iompair Éireann, PSV Circle/Omnibus Soc/TMS, 1965
Fleet History PI 5, NIRTB/UTA/Ulsterbus, PSV Circle/Omnibus Soc, 1972
Greer, PE, *Road Versus Rail*, PRON, 1973
Hunter, RA, Ludgate, RC and Richardson, J, *Gone But Not Forgotten*, ITT/ RPSI, 1979
Jenkinson, D, *British Railway Carriages of the 20th Century Volume 1*, Guild Publishing/Patrick Stephens, 1998

_____ *ibid, Volume 2*, Guild Publishing/Patrick Stephens, 1990

Jenkinson, Keth A, *Exiles In Ulster*, Autobus Review/ITT, 1990

Johnston, Norman, *The Great Northern Railway in Co Tyrone*, West Tyrone Hist Soc, 1991

_____ *Locomotives of the GNRI*, Colourpoint Books, 1999

_____and Friel, Charles, *Fermanagh's Railways*, Colourpoint Books, 1998

Jones, Peter, *Irish Railways Traction and Travel*, Metro Enterprises, 1994

Kennedy, ML and McNeill, DB, *Early Bus Services in Ulster*, IIS (QUB)/UFTM, 1997

Kidner, RW, *Narrow Gauge Railways of Ireland*, Oakwood Press, 1971

Kilroy, James, *Howth and Her Trams*, Fingal Books, 1986

_____ *Irish Trams*, Colourpoint Books, 1996

_____ *Trams to the Hill of Howth*, Colourpoint Books, 1998

McCutcheon, Alan, *Railway History in Pictures, Ireland, Volume 1*, David and Charles, 1969

_____ *ibid, Volume 2*, David and Charles, 1970

McCutcheon,WA, *The Industrial Archaeology of Northern Ireland*, HMSO, 1980

McNeill, DB, *Ulster Tramways and Light Railways*, Ulster Museum/Transport Museum, 1966

McQuillan, Jack, *The Railway Town*, Dundalgan Press, 1993

Maybin, JM, *Belfast Corporation Tramways*, Light Rail Transit Association, nd

Middlemas, Tom, *Irish Standard Gauge Railways*, David and Charles, 1981

Millar, GI, *Fifty Years of Public Service*, Ulsterbus, nd

Morton,RG, *Standard Gauge Railways in the North of Ireland*, Ulster Museum/Transport Museum, 1962

Murray, Kevin, *The Great Northern Railway (Ireland), Past, Present and Future*, Great Northern Railway, 1944

O'Riain, Michael, *On the Move*, Gill and Macmillan, 1995

Patterson, EM, *The Belfast and County Down Railway*, David and Charles, Associates, 1982

_____ *The Castlederg and Victoria Bridge Tramway*, Colourpoint Books, 1998

_____ *The Great Northern Railway of Ireland*, The Oakwood Press, 1962

_____ *The Lough Swilly Railway*, David and Charles, 1964

Prideaux, JDCA, *The Irish Narrow Gauge Railway*, David and Charles, 1981

Robb, William, *A History of Northern Ireland Railways*, William Robb, 1982

Robotham, Robert, *Last Years of the Wee Donegal*, Colourpoint Books, 1998

Rowlands and McGrath, *The Dingle Train*, Plateway Press, 1996

Rowledge, JWP, *Irish Steam Locomotive Register*, Irish Traction Group, 1993

Shepherd, Ernie, *The Midland Great Western Railway of Ireland*, Midland Publishing, 1994

Shepherd and Beesley, Dublin South Eastern Railway, Midland Publishing, 1998

Sprinks, NW, *Sligo, Leitrim and Northern Counties Railway*, IRRS, 1970

Taylor, Patrick, *The West Clare Railway*, Plateway Press, 1994

Whitehouse, P and St J Thomas, D, *LMS 150*, Guild Publishers, 1987

Twells, HN, *LMS Miscellany, Volume 1*, OPC, 1982

_____ *ibid, Volume 2*, OPC, 1984

Thompson, KM, *Lough Swilly's Fifty Years*, Irish Transport Trust, 1979

World Railways 1950-51, Sampson Low, Marston & Co Ltd, 1951

Unpublished Paper

McCutcheon, WA, 'Transport: 1820-1914'

Web Sites

Bus Éireann	www.buseireann.ie
Dublin Bus	www.dublinbus.ie
Iarnrod Éireann	www,irishrail.ie
Irish Bus	www.bustravel.net/ireland
Irish Railway Record Society	www.irrs.ie
Irish Transport Trust	www,irishtransporttrust.freeserve.co.uk
Luas light railway system	www.lrta.org
Translink	www.translink.co.uk

Index
(References in bold refer to illustrations.)

Manufacturers

Key personalities

Transport Companies

Belfast & Co Down Rly (BCDR), 14, 18-19, 21,53-55, **54**, 61,64, **68**, 69, 72, 75, 92, 132-33, **136**, 137-39, **138**, 141-42, 162, 167, 169, **174**, 182

Belfast & Northern Counties Rly (BNCR), 69, 129, 133-135, **133**, 139

Belfast Corporation Transport Dept, 17, 55, 62, 104, 150

Belfast Omnibus Co (BOC), **15**, 17-19, 139-42, **140**, 146, 149, 162-63, 167, 173

Belfast Street Tramway Co, 68, 135

Bus Átha Cliath (Dublin Bus), 9, 40-41, 43, 48-49, 51, 130, 156, 158, 161-62

Bus Éireann (Irish Bus), 9, 41, 43, 47-48, 50, 124, 130, 157-58, 162

Castlederg & Victoria Bridge Tramway (CVBT), 80

Catherwood, HMS, 15, 17-19, 61, 139, 151-52, 155, 157, 162-63

Cavan & Leitrim Rly, 80, 112, 160, 175

Central Omnibus Co, 156

Citybus, 9, 59, 61-63, 65, **127-28**, 130, **169-70**, 171-73

Clogher Valley Rly (CVR), 10, 61, 77, 81, 85-87, **87**

Clontarf & Hill of Howth Tramway, 70

Córas Iompair Éireann (CIÉ), 8-10, 14, 20, 22-24, 26-34, **28**, **31**, 33, 36, 38-50, **39**, 55, 78, 80, 84, 92, 96-97, 100-04, 109, 111-12, **122**, **124**, 130, **144-45**, 144-47, 149-50, 153-62, **155**, **158-59**, 176

County Donegal Rly (CDR), 53, 79-82, **79**, **81-82**, 85-89, **85- 87**, 97,101, 104, 107, **115**, **123**, 146-49, 156, 174, **175**, 177

Devenish Carriage & Wagon Co, 165

Dublin & Blessington Steam Tramway (DBST), 78, **82**, 174

Dublin & Kingstown Rly (DKR), 104, 130

Dublin & South Eastern Rly (DSER), 12, 73, **104**

Dublin United Tramways Co, 149, 157

Dublin United Transport Co, 158

Dublin Wexford & Wicklow Rly (DWWR), 66, 72, **73**

Erne Bus Services, 12, 18, 53, 165

General Omnibus Co, 157-58

Great Northern Railway Board (GNRB), 26-27, 55-57, 92, 146, 151, 153-54, 160

Great Northern Railway (Ireland) (GNR), 14, 16, 18, 21, 23-24, **24**, 26-27, **29**, 52, 55-58, 66-67, 69-72, **71**, 78, 81, 85-86, 88, 90-92, **91**, 96-97, 105, 107, 108-12, **110**, 141-42, 144, 146-47, 151-57, **152-55**, 160, 162, 167, 169, **174-75**, 174-76

Great Southern & Western Rly (GSWR), 12,46, 66-67, **67**, 129, **135**, 136, 174

Great Southern Railways (GSR), 8, 10-14, **13**, 22, 40, 72-75-76, **75**, 80, **83**, 95-97, 111-12, 131, 136, 142-46, **143-44**, 149, 152, 157-59, 165

Irish Omnibus Co (IOC), 10-14, **12**, 23, 40, 142-46, **143**, 150-52, 158, 165

London Midland & Scottish Rly (NCC), **7**, 14, 16, 18-19, 21, 53, 58, 73-75, **74**, 78, 86, 88, 92-94, **93-94**, 98, 104-06, **106**, **117**, **120**, 133-35, 139-42, 150, 162, **175**, 175-76,

Londonderry & Lough Swilly Rly (LLSR), **9**, 13, 17, 23, 53, 81, 97, 104, **121-22**, 146-50, **148**, **150**, 156, 161, 171, 176-77

McNeill, Henry, (Larne), 135, 139, 141

Midland Great Western Rly (MGWR), **6**, 12, **13**, 41, 79- 80, 97, 104, 129, 130-32, **131**, 143,

Midland Rly (NCC), 69, **70**, 73, 105, 107, **134**, 139

Norton of Kilkeel, 132-33

Tralee & Dingle Rly (TDR), 13, **80**, 112

General